Soil Erosion - Rainfall Erosivity and Risk Assessment

Edited by Vlassios Hrissanthou and Konstantinos Kaffas

Published in London, United Kingdom

IntechOpen

Supporting open minds since 2005

Soil Erosion - Rainfall Erosivity and Risk Assessment
http://dx.doi.org/10.5772/intechopen.74884
Edited by Vlassios Hrissanthou and Konstantinos Kaffas

Contributors
Yafei Jia, Rabii El Gaatib, Konstantinos Vantas, Epaminondas Sidiropoulos, Chris Evangelides, Rachid Nedjai, Mario Al Sayah, Chadi Abdallah, Michel Khouri, Talal Darwish, François Pinet, Vlassios Hrissanthou, Konstantinos Kaffas

First published in London, United Kingdom, 2019 by IntechOpen
IntechOpen is the global imprint of INTECHOPEN LIMITED, registered in England and Wales, registration number: 11086078, 7th floor, 10 Lower Thames Street, London, EC3R 6AF, United Kingdom
Printed in Croatia

British Library Cataloguing-in-Publication Data
A catalogue record for this book is available from the British Library

Additional hard and PDF copies can be obtained from orders@intechopen.com

Soil Erosion - Rainfall Erosivity and Risk Assessment
Edited by Vlassios Hrissanthou and Konstantinos Kaffas
p. cm.
Print ISBN 978-1-78985-195-3
Online ISBN 978-1-78985-196-0
eBook (PDF) ISBN 978-1-78985-581-4

For EU product safety concerns:
IN TECH d.o.o., Prolaz Marije Krucifikse Kozulić 3, 51000 Rijeka, Croatia, info@intechopen.com or visit our website at intechopen.com.

Meet the editors

Dr. Ing. Vlassios Hrissanthou is an Emeritus Professor in the Civil Engineering Department of the Democritus University of Thrace, Xanthi, Greece. He studied Civil Engineering at the Aristotle University of Thessaloniki, Greece. He then undertook postgraduate and doctoral studies on hydrology and hydraulic structures at the Karlsruhe Institute of Technology, Germany. Subsequently, he completed a postdoctoral study on hydraulics and hydraulic structures at the University of the Armed Forces Munich, Germany. His teaching work includes seven graduate and six postgraduate study courses on hydraulics, hydrology and hydraulic structures. He is the author and co-author of 140 publications in scientific journals, conference proceedings, and book chapters in English, Greek, and German. He has reviewed numerous publications for 42 international scientific journals.

Dr. Eng. Konstantinos Kaffas is a researcher in the Faculty of Science and Technology of the Free University of Bozen–Bolzano, Bolzano, Italy. He received his diploma in Environmental Engineering from the Environmental Engineering Department of the Democritus University of Thrace, Xanthi, Greece, in 2007. Subsequently, he pursued postgraduate studies (M.Sc.) in hydraulic engineering and water resources management works in the Civil Engineering Department of the latter university. He obtained his PhD on hydrology, soil erosion and sediment transport at the basin scale, from the same department. His current scientific interests are focused on meteorological and hydrological forecasting and downscaling. He has authored a large number of publications and he has reviewed articles for over ten international scientific journals.

Contents

Preface

This book is on soil erosion caused by rainfall and runoff at the basin scale. Why is the study of soil erosion significant? Soil erosion products are transported by runoff into the streams of a basin and through the streams to the basin outlet, which may also be a coastal zone or the inlet of a natural or artificial lake. The transport of large amounts of eroded soil by runoff, and consequently the transport of large amounts of suspended sediment in the streams, is mainly dependent on the frequency and intensity of rainfall events. Removal of fertile soil and acceleration of reservoir sedimentation are some additional unfavorable effects of soil erosion. Soil erosion modeling contributes to the quantification of eroded soil and informs actions against soil erosion.

This book contains three sections. In the first section, soil erosion is defined, the stages of soil erosion are described, factors influencing soil erodibility are discussed, and severe soil erosion consequences are reported. Additionally, the historical evolution of soil erosion models is briefly presented. In the first part of the second section, the coefficient of rainfall erosivity in the Universal Soil Loss Equation (USLE) is estimated on the basis of pluviograph records on the one hand, and cumulative rainfall depths by means of empirical equations and machine learning methods on the other hand. Rainfall data from 84 meteorological stations in Greece are used. In the second part of the second section, a physically-based, hydrodynamic, finite element model is analytically described for the computation of surface runoff, that causes soil erosion and contributes to channel flows. The model is applied to a low-relief agricultural basin in the Mississippi River alluvial plain. In the first part of the third section, soil erosion risk is assessed under different geographical, topographical, climatological, and land occupation/management scenarios, in the French Claise and Lebanese Nahr Ibrahim basins. In the second part of the third section, the erosion risk management in the OuedBeht basin (Morocco) is evaluated, and the delimitation of the areas requiring priority planning is achieved.

Vlassios Hrissanthou
Emeritus Professor,
Section of Hydraulic Engineering,
Department of Civil Engineering,
Democritus University of Thrace,
Greece

Konstantinos Kaffas
Free University of Bozen-Bolzano,
Italy

Section 1

Introduction to Soil Erosion Problem

Chapter 1

Introductory Chapter: Soil Erosion at a Glance

Konstantinos Kaffas and Vlassios Hrissanthou

1. Introduction

Wind and precipitation are the two weather elements prevailing as the generating causes of soil erosion, inducing the so-called wind erosion and water erosion. While erosion by wind is internationally termed wind erosion, erosion caused by water can be found by a variety of definitions, such as water erosion, sheet erosion, surface erosion, rill erosion, interrill erosion, land erosion, and soil erosion. Most frequently, it is referred to as soil erosion and as such will be denoted in the present chapter. Despite the fact that there is ample literature on both erosion types and though wind erosion is a major environmental problem, especially in large open areas, with sparse vegetation and loose soils, modelers have put appreciably greater effort on studying soil erosion as a hydrologically-driven magnitude. This short review is dedicated to soil erosion, aiming to report some key information on its background and modeling.

1.1 Definition of the problem

Soil erosion, globally recognized as the main cause of land degradation, is the physical phenomenon, triggered by rainfall, during which soil particles are detached from the soil mass and washed downslope by surface runoff. Sediment is detached from soil surface both by the raindrop impact and shearing force of flowing water [1]. Thus, soil erosion is mainly due to rainfall and runoff. Erosion due to rainfall, also known as splash erosion, is the first stage in the water erosion process. At this stage, raindrops act like little bombs detaching soil particles and destroying soil structure [2]. Subsequently, the detached soil particles are carried away by the flowing water and the soil can be further eroded, depending on the runoff's transport capacity. At this point, reasonable questions are raised: What happens to the soil surface when soil is continuously removed, due to erosion by wind, rainfall and runoff? Would not that lead to constant drop of soil surface level? And what effect would that have to plants, ecosystems, and humans? The answer is that as soil gets eroded and removed, it is also formed by the physical, chemical and biological weathering of rocks. Ultimately, it could be stated that there is a permanent soil loss when soil is removed at rates greater than the ones it is formed.

Among the obvious consequences are soil loss and changes in the land surface morphology; yet, the implications go much deeper. It is well documented that soil erosion leads to decline of soil fertility and to considerable loss of productive cultivated and arable land or even to desertification, with serious socioeconomic effects [3–8]. A degradation of physical properties of soil involves a decline in soil structure resulting to an increase in bulk density, decrease in total macroporosity, reduction in infiltration, and increase in surface runoff and, finally, in aggravation of soil erosion by water [9]. Apart from the effects on soil fertility, soil development, soil

IntechOpen

degradation, and soil diversity [10], an eroded and fractured soil can ease the deep percolation of water and sediments to the aquifer. Moreover, surface runoff washes down toward the stream network, carrying sediments that have been previously abstracted from the soil. As stated by Unger and McCalla [11], water erosion is a major contributor to water pollution. Hence, groundwater and surface water contamination by influx of rain water and sediments, which can be carriers of polluting factors, is a further side effect, especially in agricultural basins where fertilizers and pesticides are in use. Generally, soil erosion is directly connected to a series of environmental issues, such as problems with the vegetation growth, increase of soil acidity levels, muddy floods, etc.

It would not be an exaggeration to say that the effects of soil erosion are more evident in the adjacent fluvial systems than in the soil surface itself. The greatest part (more than 90%, many times) of the instream sediment derives from sediment inflow as the product of rainfall and runoff erosion. Thus, a river, in effect, can be considered a body of flowing sediments as much as one of flowing water [12]. Depending on the hydraulic conditions and the sediment transport capacity of streamflow, this sediment gets deposited or/and transported downstream, unceasingly embroidering the morphological profile of rivers and streams. The influx of erosion yields to the streams has positive and negative impacts. Aside from what was discussed above regarding potential contaminations, sediments are also carriers of nutritional factors, necessary for the thriving of riparian and fluvial ecosystems. A balanced amount of deposition provides the appropriate grounds of spawning for fish and macroinvertebrates. Contrarily, excessive sedimentation can cause changes in faunal assemblages, the decline of macrophyte growth, and the clogging of spawning gravel [13] or even, effectively, ravage their natural habitats. Excessive deposition can cause increase of flood events, by diminishing the cross-sectional areas. High concentrations of sediment—as a product of soil erosion—in rivers can lead to degradation of water quality, which in turn would result in an increase of water treatment costs. Sediment transport, also, greatly affects the morphology of the shoreline and the coastal zones. According to Samaras and Koutitas [14], coastal areas are subject to "pressures" from upstream watersheds in terms of sediment transport.

Soil erosion, implicitly, takes a toll on hydraulic structures, such as reservoirs and hydroelectric schemes. Sediments constitute—even today—the worst implication associated with dams, due to excessive sedimentation which leads to a considerable storage loss. Even when sediment flushing is a viable solution for recovering and maintaining storage capacity of small- to medium-sized hydropower reservoirs, observations have documented significant environmental damage due to sediment release downstream [15, 16]. According to Cui et al. [17], the accumulation of fine sediment in reservoirs and the potential impact of sediment flushing constitute even a cause for dams to be removed in some cases. It is true that in several cases, dams are decommissioned or even abandoned due to sedimentation (Nizam Sagar dam, Katteri dam, Bhakra dam (India), Sanmenxia dam (China), Peligre dam (Haiti), Melton dam, Umberumberka dam (Australia), and others [18]). This is, obviously, not attributed to manufacturing defects but to failed prediction of the sediment discharge at the location of the dam, prior to its construction, in other words, to the underestimation of soil erosion and sediment discharge of basins drained by dams.

1.2 Factors influencing soil erosion

In order to assess if and under what conditions erosion will take place, a critical question has to be answered: are all soils equally prone to erosion?

As to their susceptibility to erosion, soils can be placed into a spectrum ranging from erodible to non-erodible. Erodible are usually characterized non-cohesive soils with little or no resistance to erosion, while non-erodible are soils notably less susceptible to erosion. This characterization is made on the basis of the physical properties of soil, alone, regardless any exogenous factors like land cover, land use, or support practices. Studies on the physical properties of soil have shown that soil texture (sand, silt, clay content) and organic matter play an important role to soil erodibility [19–23]. According to Wischmeier and Mannering [24], a soil's inherent erodibility is a complex property dependent both on its infiltration capacity and its capacity to resist detachment by rainfall and transport by runoff. Thus, the effect of soil characteristics can be observed in two consecutive stages, first being the endurance of soil to the raindrop impact and its resistance to detachment. The more concrete is the structure of the soil, the more armored it is against splash erosion. The second stage initiates with surface runoff, when the intensity of rainfall exceeds the infiltration capacity of the soil. Most relevant studies point out silt as the main culprit for soil susceptibility to erosion. In fact, the effect of the silt content in a soil is such that it can itself be a regulating factor of the soil's erodibility. Wischmeier and Mannering [24] report that a soil type becomes less erodible with decrease in silt fraction, regardless of whether the corresponding increase is in the sand fraction or the clay fraction. Generally, silty and sandy soils with low content in clay and organic matter are known to be more prone to erosion [23–25]. For a better comprehension of soils' texture, FAO's World Reference Base for Soil Resources [26] provides particle classes, according to their size (**Table 1**).

Despite being well-aggregated, silty soils suffer a collapse of their aggregations when wetted, allowing the non-aggregated fine particles to be easily transported by runoff [22]. Sandy soils are, also, susceptible to detachment due to their low cohesion, but their high permeability to water, resulting in low runoff rates, in combination with their large size and density, makes it difficult to transport by runoff. Clayey soils are characterized by high cohesion and low infiltration rates; they are very resisting to detachment but are easily transported, once detached from the soil body.

Despite what was discussed above, soil erosion processes are characterized by even greater complexity. An ensemble of additional parameters, such as land cover type, land use practices, weather conditions, etc., influences soil erosion at a large extent. As shown by Morin and Benyamini [27], the antecedent moisture conditions, as well as the duration and intensity of rainfall, play an important role and cannot be ignored. It is well-known that the denser the land cover and canopy, the more the raindrop impact, and thus the erosive force of rainfall is contained. Mohammad and Adam [28] support that the lowest runoff and soil erosion rates are associated with the forest and with natural vegetation. The effect of land cover

Soil texture	Diameter limits (mm)
Very coarse sand	1.25–2.00
Coarse sand	0.63–1.25
Medium sand	0.20–0.63
Fine sand	0.125–0.20
Very fine sand	0.063–0.125
Coarse silt	0.02–0.063
Fine silt	0.002–0.02
Clay	<0.002

Table 1.
Particle size classes (WRBSR-FAO).

and land use practices on soil erosion at the basin scale has been well documented—among others—in [28–31].

As stated—very early—by Middleton [19], all soils are somewhat susceptible to erosion by runoff water. Thus, reliable information on soil erosion rates is an essential prerequisite for the design of targeted erosion and sediment control strategies [32].

2. Review of literature

What is mentioned in the previous sections dictates the necessity for soil erosion quantification and highlights soil erosion modeling as the utmost vital action taken, in the context of an integrated management at the basin scale.

Soil degradation by accelerated erosion is a serious problem and will remain so during the twenty-first century. Soil erosion prediction and assessment have been a challenge to researchers since the 1930s, and several models have since been developed [33]. However, the treatment of soil erosion in the form of soil conservation plans has made its appearance long before that, in the early nineteenth century. As stated by Dotterweich [34], the first extensive essay on soil conservation known to the western world was published in Germany in 1815, while the rise of professional soil conservation occurred in the late nineteenth and early twentieth centuries. Substantially, the first decades of the nineteenth century can be considered as the outset of the profound understanding and studying of the phenomenon of soil erosion. It is remarkable that the third president of the United States, Thomas Jefferson, in one of his letters in 1813 [35], demonstrates his awareness of the on- and off-site effects of soil erosion, the role of runoff in soil erosion, and the interaction of soil conservation, hydrology, and crop production, important scientific topics today in understanding, predicting, and modeling soil erosion, 200 years later [36].

The most significant, and groundbreaking for that time, theory regarding soil erosion was introduced in 1899 by Davis. His theory, known as cycle of erosion [37], is an idealized model for stream erosion and landscape development in which stream erosion occurs in a gradual sequence of stages (young, mature, and old). During these stages, the soil surface erodes up to the point it becomes a peneplain. Davis' cycle of erosion dominated in geomorphology for more than half a century.

After 1950, the Davisian theory began to be questioned. Among those who challenged it was Chorley [38, 39] who rejected the Davisian cycle of erosion and suggested a quantitative method based on general system theory and numerical modeling. King has also tried to dispute the Davisian cycle of erosion [40, 41]; however, his theories did not manage to escape the Davisian cyclical nature. According to Bishop [42], the dissatisfaction was embodied in Strahler's [43] call for radical change and the embracing of a new approach and underpinning concepts, ultimately taking the discipline into spatial and temporal scales much reduced from the grand vision and sweeping canvas of Davis and his disciples. Strahler's call is now being heard in long-term landscape evolution as geomorphology embraces quantitative and geochemical analytical approaches to the sorts of questions that Davis sought to address [42].

Boardman [44] highlights some of the most notable and influential advances of the recent past, among which are the following: Trimble [45] with its emphasis on sediment storage and the relationship between erosion on the hillslopes and the role of the valley-bottom stream; Govers' and Poesen's [46] empirical study of rill and interrill areas; De Ploey's [47] attempt to categorize eroding western European landscapes; and Blaikie's [48] recognition that degradation occurs because of people-land relationships often involving social and economic opportunities and constraints [44].

The Universal Soil Loss Equation (USLE) [49] is one of the most significant advances in soil and water conservation in the twentieth century. It has been applied in almost all the kinds of climatic conditions and types of soils around the globe, as an individual model, while it also constitutes an important component of many models and hydromorphological softwares. Since then, there have been many parallaxes and modifications of USLE, the most known of which are the Modified Universal Soil Loss Equation (MUSLE) [50] and the Revised Universal Soil Loss Equation (RUSLE) [51].

The evolution of the soil erosion and the sediment transport modeling has consistently followed the evolution of technology. In the last few decades, there has been a hectic advancement in the domain of soil erosion modeling, as a result of the advancements in computer science. This resulted in the development of a plethora of integrated models that—in many cases—fully address the study of the hydromorphological processes. There is a wide range of integrated models that simulate the runoff, the soil erosion, and the stream sediment transport processes, on a continuous (long-term) or on an event-time basis. Some notorious examples, with a prominent position in literature, are the following: the Agricultural Nonpoint Source (AGNPS) model [52], the Chemical Runoff and Erosion from Agricultural Management Systems (CREAMS) model [53], the Areal Nonpoint Source Watershed Environment Response Simulation (ANSWERS) [54], the Soil and Water Assessment Tool (SWAT) [55, 56], the European Soil Erosion Model (EUROSEM) [57], the Hydrologic Engineering Center-Hydrologic Modeling System (HEC-HMS) [58], and the Watershed Erosion Prediction Project (WEPP) [59]. These models have been applied both stand-alone and as a part of integrated mathematical models, to model the sedimentary cycle [60–63].

Author details

Konstantinos Kaffas[1]* and Vlassios Hrissanthou[2]

1 Faculty of Science and Technology, Free University of Bozen-Bolzano, Bozen-Bolzano, Italy

2 Department of Civil Engineering, Democritus University of Thrace, Xanthi, Greece

*Address all correspondence to: konstantinos.kaffas@unibz.it

References

[1] Jain MK, Das D. Estimation of sediment yield and areas of soil erosion and deposition for watershed prioritization using GIS and remote sensing. Water Resources Management. 2010;**24**(10):2091-2112. DOI: 10.1007/s11269-009-9540-0

[2] Cotching WE. Soil Health for Farming in Tasmania. Devonport, Tasmania: Richmond Concepts and Print; 2009

[3] Lal R. Soil Erosion Research Methods. 3rd ed. Ankeny, IOWA: Soil and Water Conservation Society; 1994

[4] Koulouri M, Giourga C. Land abandonment and slope gradient as key factors of soil erosion in Mediterranean terraced lands. Catena. 2007;**69**(3):274-281. DOI: 10.1016/j.catena.2006.07.001

[5] Çetin SC, Karaca A, Haktanır K, Yildiz H. Global attention to Turkey due to desertification. Environmental Monitoring and Assessment. 2007;**128**(1-3):489-493. DOI: 10.1007/s10661-006-9342-2

[6] Panagos P, Standardi G, Borrelli P, Lugato E, Montanarella L, Bosello F. Cost of agricultural productivity loss due to soil erosion in the European Union: From direct cost evaluation approaches to the use of macroeconomic models. Land Degradation and Development. 2018;**29**(3):471-484. DOI: 10.1002/ldr.2879

[7] Novara A, Pisciotta A, Minacapilli M, Maltese A, Capodici F, Cerdà A, et al. The impact of soil erosion on soil fertility and vine vigor. A multidisciplinary approach based on field, laboratory and remote sensing approaches. Science of the Total Environment. 2018;**622**:474-480. DOI: 10.1016/j.scitotenv.2017.11.272

[8] Chalise D, Kumar L, Spalevic V, Skataric G. Estimation of sediment yield and maximum outflow using the IntErO model in the Sarada River Basin of Nepal. Water. 2019;**11**(5):952. DOI: 10.3390/w11050952

[9] Lal R, editor. Soil erosion by wind and water: Problems and prospects. In: Soil Erosion Research Methods. New York, USA: Routledge; 2017. pp. 1-10

[10] Dotterweich M. The history of soil erosion and fluvial deposits in small catchments of Central Europe: Deciphering the long-term interaction between humans and the environment—A review. Geomorphology. 2008;**101**(1-2):192-208. DOI: 10.1016/j.geomorph.2008.05.023

[11] Unger PW, McCalla TM. Conservation tillage systems. Advances in Agronomy. 1980;**33**:1-58. DOI: 10.1016/S0065-2113(08)60163-7

[12] McCully P. Silenced River: The Ecology and Politics of Large Dams. London, United Kingdom: Zed Books; 2001

[13] Graf W, Leitner P, Hanetseder I, Ittner LD, Dossi F, Hauer C. Ecological degradation of a meandering river by local channelization effects: A case study in an Austrian lowland river. Hydrobiologia. 2016;**772**(1):145-160. DOI: 10.1007/s10750-016-2653-6

[14] Samaras AG, Koutitas CG. Modeling the impact of climate change on sediment transport and morphology in coupled watershed-coast systems: A case study using an integrated approach. International Journal of Sediment Research. 2014;**29**(3):304-315. DOI: 10.1016/S1001-6279(14)60046-9

[15] Crosa G, Castelli E, Gentili G, Espa P. Effects of suspended sediments

from reservoir flushing on fish and macroinvertebrates in an alpine stream. Aquatic Sciences. 2010;**72**(1):85. DOI: 10.1007/s00027-009-0117-z

[16] Brown CB. The Control of Reservoir Silting, Miscellaneous Publication (No. 521). Washington, DC: United States Department of Agriculture; 1943

[17] Cui Y, Booth DB, Monschke J, Gentzler S, Roadifer J, Greimann B, et al. Analyses of the erosion of fine sediment deposit for a large dam-removal project: An empirical approach. International Journal of River Basin Management. 2017;**15**(1):103-114. DOI: 10.1080/15715124.2016.1247362

[18] Kaffas K, Hrissanthou V, Sevastas S. Modeling hydromorphological processes in a mountainous basin using a composite mathematical model and ArcSWAT. Catena. 2018;**162**:108-129. DOI: 10.1016/j.catena.2017.11.017

[19] Middleton HE. Properties of Soils which Influence Soil Erosion (U.S. Department of Agriculture Technical Bulletin No. 178). Washington, DC: U.S. Department of Agriculture; 1930

[20] Wischmeier WH, Johnson CB, Cross BV. Soil erodibility nomograph for farmland and construction sites. Journal of Soil and Water Conservation. 1971;2:189-193

[21] Meyer LD, Harmon WC. Susceptibility of agricultural soils to interrill erosion 1. Soil Science Society of America Journal. 1984;**48**(5):1152-1157. DOI: 10.2136/sssaj1984.03615995004800050040x

[22] Gilley JE. Erosion water-induced. In: Hillel D, editor. Encyclopedia of Soils in the Environment. Elsevier; 2005

[23] Finch S, Samuel A, Lane GP. Fertilisers and manures. In: Lockhart and Wiseman's Crop Husbandry Including Grassland. 9th ed. Woodhead Publishing/Elsevier; 2014. DOI: 10.1533/9781782423928.1.37

[24] Wischmeier WH, Mannering JV. Relation of soil properties to its erodibility. Soil Science Society of America Journal. 1969;**33**(1):131-137. DOI: 10.2136/sssaj1969.03615995003300010035x

[25] Finch S, Samuel A, Lane GP. Lockhart and Wiseman's Crop Husbandry Including Grassland. 8th ed. Woodhead Publishing/Elsevier. DOI: 10.1533/9781855736504.1.25

[26] Iuss Working Group Wrb. World reference base for soil resources 2014, update 2015: International soil classification system for naming soils and creating legends for soil maps. World Soil Resources Reports. 2015;**106**:192

[27] Morin J, Benyamini Y. Rainfall infiltration into bare soils. Water Resources Research. 1977;**13**(5):813-817. DOI: 10.1029/WR013i005p00813

[28] Mohammad AG, Adam MA. The impact of vegetative cover type on runoff and soil erosion under different land uses. Catena. 2010;**81**(2):97-103. DOI: 10.1016/j.catena.2010.01.008

[29] Martínez-Casasnovas JA, Sánchez-Bosch I. Impact assessment of changes in land use/conservation practices on soil erosion in the Penedès–Anoia vineyard region (NE Spain). Soil and Tillage Research. 2000;**57**(1-2):101-106. DOI: 10.1016/S0167-1987(00)00142-2

[30] Panagos P, Borrelli P, Meusburger K, van der Zanden EH, Poesen J, Alewell C. Modelling the effect of support practices (P-factor) on the reduction of soil erosion by water at European scale. Environmental Science & Policy. 2015;**51**:23-34. DOI: 10.1016/j.envsci.2015.03.012

[31] Turner BL, Fuhrer J, Wuellner M, Menendez HM, Dunn BH, Gates R. Scientific case studies in land-use driven soil erosion in the Central United States: Why soil potential and risk concepts should be included in the principles of soil health. International Soil and Water Conservation Research. 2018;**6**(1):63-78. DOI: 10.1016/j.iswcr.2017.12.004

[32] Collins AL, Walling DE, Sichingabula HM, Leeks GJL. Using 137 Cs measurements to quantify soil erosion and redistribution rates for areas under different land use in the Upper Kaleya River Basin, Southern Zambia. Geoderma. 2001;**104**(3):299-323. DOI: 10.1016/S0016-7061(01)00087-8

[33] Lal R. Soil degradation by erosion. Land Degradation and Development. 2001;**12**(6):519-539. DOI: 10.1002/ldr.472

[34] Dotterweich M. The history of human-induced soil erosion: Geomorphic legacies, early descriptions and research, and the development of soil conservation—A global synopsis. Geomorphology. 2013;**201**:1-34. DOI: 10.1016/j.geomorph.2013.07.021

[35] Jefferson T. In: Betts EM, editor. Letter to CW Peale, Apr. 17, 1813. Thomas Jefferson's Garden Book. Philadelphia, PA: American Philosophical Society; 1944. p. 509

[36] Laflen JM, Flanagan DC. The development of US soil erosion prediction and modeling. International Soil and Water Conservation Research. 2013;**1**(2):1-11. DOI: 10.1016/S2095-6339(15)30034-4

[37] Davis WM. The geographical cycle. The Geographical Journal. 1899;**14**(5):481-504. DOI: 10.2307/1774538

[38] Chorley RJ. A re-evaluation of the geomorphic system of WM Davis. Frontiers in Geographical Teaching. 1965:21-38

[39] Chorley RJ, Beckinsale RP, Dunn AJ. The History of the Study of Landforms or the Development of Geomorphology: The Life and Work of William Morris Davis. Vol. 2. Psychology Press; 1973

[40] King LC. The study of the world's plainlands: A new approach in geomorphology. Quarterly Journal of the Geological Society. 1950;**106**(1-4):101-131. DOI: 10.1144/GSL.JGS.1950.106.01-04.06

[41] King LC. Canons of landscape evolution. Geological Society of America Bulletin. 1953;**64**(7):721-752. DOI: 10.1130/0016-7606(1953)64[721:COLE]2.0.CO;2

[42] Bishop P. Long-term landscape evolution: Linking tectonics and surface processes. Earth Surface Processes and Landforms. 2007;**32**(3):329-365. DOI: 10.1002/esp.1493

[43] Strahler AN. Dynamic basis of geomorphology. Geological Society of America Bulletin. 1952;**63**(9):923-938. DOI: 10.1130/0016-7606(1952)63[923:DBOG]2.0.CO;2

[44] Boardman J. Soil erosion science: Reflections on the limitations of current approaches. Catena. 2006;**68**(2):73-86. DOI: 10.1016/j.catena.2006.03.007

[45] Trimble SW. A sediment budget for Coon Creek basin in the Driftless area, Wisconsin, 1853-1977. American Journal of Science. 1983;**283**(5):454-474. DOI: 10.2475/ajs.283.5.454

[46] Govers G, Poesen J. Assessment of the interrill and rill contributions to total soil loss from an upland field plot. Geomorphology. 1988;**1**(4):343-354. DOI: 10.1016/0169-555X(88)90006-2

[47] De Ploey J. Erosional systems and perspectives for erosion control in European loess areas. In: Schwertmann U, Rickson RJ, Auerswald K, editors. Soil Erosion Protection Measures in Europe. Germany: Catena Verlag; 1989. pp. 93-102

[48] Blaikie PM. The Political Economy of Soil Erosion in Developing Countries. Harlow, Essex, England: Longman Scientific & Technical/Wiley; 1985

[49] Wischmeier WH, Smith DD. Predicting rainfall-erosion losses from cropland east of the Rocky Mountains. In: Agriculture Handbook No. 282, U.S. Department of Agriculture, Washington, DC; 1965

[50] Williams JR. Sediment-yield prediction with universal soil loss equation using runoff energy factor. In: Present and Prospective Technology for Predicting Sediment Yields and Sources. Proceedings of the Sediment Yield Workshop; Oxford, MS; 1975. pp. 244-252

[51] Renard KG, Foster GR, Weesies GA, McCool DK, Yoder DC. Predicting Soil Erosion by Water: A Guide to Conservation Planning with the Revised Universal Soil Loss Equation (RUSLE) (Vol. 703). Washington DC: US Government Printing Office; 1996

[52] Young RA, Onstad CA, Bosch DD, Anderson WP. AGNPS, Agricultural Nonpoint Source Pollution. A Watershed Analysis Tool. Conservation Research Report 35. Washington, DC: US Department of Agriculture; 1987

[53] Knisel WG. CREAMS: A Field-Scale Model for Chemicals, Runoff, and Erosion from Agricultural Management System. Conservation Research Report No. 26. Washington, DC: USDASEA; 1980

[54] Beasley DB, Huggins LF, Monke EJ. ANSWERS: A model for watershed planning. Transactions of ASAE. 1980;**23**(4):938-944. DOI: 10.13031/2013.34692

[55] Neitsch SL, Arnold JG, Kiniry JR, Williams JR. Soil and Water Assessment Tool Theoretical Documentation Version 2009. College Station, Texas, USA: Texas Water Resources Institute, Texas A&M University; 2011

[56] Arnold JG, Kiniry JR, Srinivasan R, Williams JR, Haney EB, Neitsch SL. Input/Output Documentation Version 2012. Texas, USA: Texas Water Resources Institute; 2012

[57] Morgan RPC, Quinton JN, Smith RE, Govers G, Poesen JWA, Auerswald K, et al. The European Soil Erosion Model EUROSEM: A dynamic approach for predicting sediment transport from fields and small catchments. Earth Surface Processes and Landforms. 1998;**23**(6):527-544. DOI: 10.1002/(SICI)1096-9837(199806)

[58] United States Army Corps of Engineers (USACE). Hydrologic Modeling System HEC-HMS, User's Manual, Version 3.5. Davis, California, USA: Hydrologic Engineering Center; 2010

[59] Laflen JM, Lane LJ, Foster GR. WEPP: A new generation of erosion prediction technology. Journal of Soil and Water Conservation. 1991;**46**(1):34-38

[60] Yesuf HM, Assen M, Alamirew T, Melesse AM. Modeling of sediment yield in Maybar gauged watershed using SWAT, Northeast Ethiopia. Catena. 2015;**127**:191-205. DOI: 10.1016/j.catena.2014.12.032

[61] Pak J, Fleming M, Scharffenberg W, Ely P. Soil erosion and sediment yield modeling with the hydrologic modeling system (HEC-HMS). In: World Environmental and Water Resources

Congress 2008. Honolulu, Hawaii: Ahupua'A; 2008. pp. 1-10

[62] Kaffas K, Hrissanthou V. Computation of hourly sediment discharges and annual sediment yields by means of two soil erosion models in a mountainous basin. International Journal of River Basin Management. 2019;**17**(1):63-77. DOI: 10.1080/15715124.2017.1402777

[63] Mati BM, Morgan RPC, Quinton JN. Soil erosion modelling with EUROSEM at Embori and Mukogodo catchments, Kenya. Earth Surface Processes and Landforms. 2006;**31**(5):579-588. DOI: 10.1002/esp.1347

Section 2

Computation of Rainfall Erosivity and Surface Runoff

Chapter 2

Rainfall Erosivity and Its Estimation: Conventional and Machine Learning Methods

Konstantinos Vantas, Epaminondas Sidiropoulos and Chris Evangelides

Abstract

Rainfall erosivity concerns the ability of rainfall to cause erosion on the surface of the earth. The difficulty in modeling the distribution, the size, and the terminal velocity of raindrops in relation to the detachment of soil particles led to the use of more tractable rainfall indices. Thus, in the universal soil loss equation (USLE), the coefficient of rainfall erosivity, R, was introduced. This coefficient is based on the product of the rainfall kinetic energy of a storm and its maximum 30-minute intensity. An important problem in the application of USLE and its revisions in various parts of the world concerns the computation of R, which requires pluviograph records with a length of at least 20 years. For this reason, empirical equations have been developed that are based on coarser rainfall data, such as daily, monthly, or yearly, which are available on larger spatial and temporal extents. However, the lack of denser data is dealt more effectively by means of machine learning methods. Computational systems for this purpose were recently developed based on feed-forward neural networks, yielding significantly better results.

Keywords: rainfall, erosivity, machine learning, erosivity density, universal soil loss equation, nonlinear regression, neural networks

1. Introduction

Soil erosion is the detachment, transport, and deposition of soil particles, and it takes place in the course of one or more processes. These processes are natural, such as rainfall with the energy that it carries, surface water runoff, wind, and gravity. The observed processes may be combined with human activities and works, such as intensive cultivations, deforestation, construction of public works, and mining activities [1].

Earth terrain is strongly influenced by erosion, which progresses in geological time, in case it is a purely natural process. Otherwise, when human activities and works are involved, the phenomenon of erosion is accelerated. This accelerated erosion may cause uncontrollable soil loss with negative consequences for human health, the natural ecosystems, the climate, as well as the economy.

The universal soil loss equation (USLE) with its revisions (RUSLE, RUSLE2) can be used to predict the average rate of soil erosion by grouping the numerous parameters that affect soil loss into a set of factors and is the prediction equation most widely used in the world [2]. According to Nearing et al. [3], "Soil loss refers to the amount of sediment that reaches the end of a specified area on a hillslope that is experiencing net loss of soil by water erosion." Development of soil loss equations began in 1940, and the universal soil loss equation was developed at the National Runoff and Soil Loss Data Center established in 1954 by the Science and Education Administration [4].

The USLE represents an erosion model developed for the prediction of soil losses in an average long-term sense. It is based on the knowledge of the physical characteristics of the field area under study, along with the prevailing cropping and management system. USLE has been widely tested in field conditions, and therefore its validity has been established.

USLE consists of the product of six factors, whose numerical values can be specified depending on a particular location. There is a considerable variation in the resulting erosion values, if observations are limited within short periods. However, long-term averages correspond more satisfactorily to predictions.

The soil loss equation is

$$A = R \cdot K \cdot L \cdot S \cdot C \cdot P \tag{1}$$

where the meaning of the symbols is given in [5] exactly as follows:

- *A is the computed soil loss per unit area, expressed in the units selected for K and for the period selected for R.*
- *R, the rainfall and runoff factor, is the number of rainfall erosion index units, plus a factor for runoff from snowmelt or applied water where such runoff is significant.*
- *K, the soil erodibility factor, is the soil loss rate per erosion index unit for a specified soil as measured on a unit plot, which is defined as a 72.6-ft length of uniform 9% slope continuously in clean-tilled fallow.*
- *L, the slope length factor, is the ratio of soil loss from the field slope length to that from a 72.6-ft length under identical conditions.*
- *S, the slope-steepness factor, is the ratio of soil loss from the field slope gradient to that from a 9% slope under otherwise identical conditions.*
- *C, the cover and management factor, is the ratio of soil loss from an area with specified cover and management to that from an identical area in tilled continuous fallow.*
- *P, the support practice factor, is the ratio of soil loss with a support practice like contouring, strip cropping, or terracing to that with straight-row farming up and down the slope.*

Out of the six factors, only two have units, the erosivity R and the erodibility K. The remaining four, i.e., slope steepness, slope length, cropping management, and components of control practice, are dimensionless factors, because they represent ratios in relation to the unit plot [4].

Regarding R, it needs to be observed that its units stem from its definition as the energy multiplied by maximum 30-minute intensity term. In contrast, the erodibility factor values were determined empirically by calibration against measured erosion data. This is important due to the fact that if there is a change on the definition of the R factor, then the values of the K factor should be recalculated [3].

2. Computation of the *R* factor

The *R* factor is computed for time periods greater than 20 years, so that wet or dry rainfall periods will be incorporated, eliminating any bias. It is set equal to the average of the sum of the erosivity values for every year's rainfalls. The *R* coefficient is defined as the product of the kinetic energy of an erosive rainfall event by the maximum intensity of a 30-minute duration rainfall, during the rainfall event:

$$R = \frac{1}{n} \sum_{j=1}^{n} \sum_{k=1}^{m_j} (EI_{30})_k \tag{2}$$

where n is the number of years in the record, m_j the number of erosive events during year j, and EI_{30} (MJ mm ha^{-1} h^{-1}) the rainfall erosivity for event k.

The erosivity of a particular event is

$$EI_{30} = \left(\sum_{r=1}^{m} e_r \cdot v_r \right) \cdot I_{30} \tag{3}$$

where e_r is the kinetic energy per unit of rainfall (MJ ha^{-1} mm^{-1}); v_r is the rainfall depth (mm) for the time interval r of the hyetograph, which has been divided into $r = 1, 2, \ldots, m$ subintervals; and I_{30} is the maximum rainfall intensity for a 30-minute duration.

For the computation of e_r, numerous empirical relations (**Figure 1**) involving rain intensity have been proposed [6]. Thus, in USLE, the empirical relation of Wischmeier and Smith [4] was used:

$$e_r = 0.119 + 0.0873 log_{10}(i) \tag{4}$$

with the upper limit of 0.283 MJ ha^{-1} mm^{-1} if i>76 mm h^{-1}, where i is rainfall intensity.

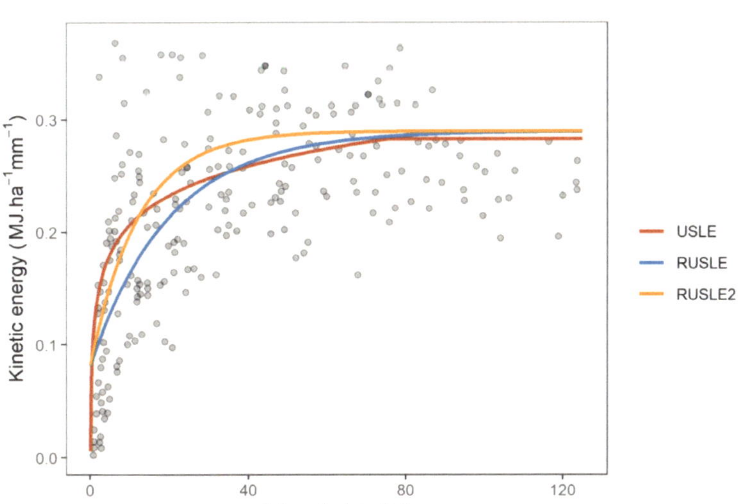

Figure 1.
The three different kinetic energy equations used in USLE and its revisions. The points are actual data, showing the relation between intensity and the kinetic energy of the rainfall, converted to SI units and coming from Haan et al. [61].

Later, in RUSLE [5], the exponential relation of Brown and Foster [7] was used:

$$e_r = 0.29\left(1 - 0.72\ e^{-0.05i}\right) \tag{5}$$

Upon comparison of the equations used in USLE and RUSLE, McGregor and Mutchler [8] found out that for rain intensities up to 35 mm/h, Eq. (5) yields results by 12% less than those of Eq. (4), and they proposed a modification in the value 0.05 that controls the rate of change of e_r with i. Thus, in RUSLE, the following relation has been adopted:

$$e_r = 0.29\left(1 - 0.72\ e^{-0.082i}\right) \tag{6}$$

Eq. (5) was developed for an application concerning reordered rainfall intensity data and not natural rainfall data [6]. The systematic underestimation of R has been shown in many other studies [3, 7–9], so Eq. (5) should not be used for calculations. The rules that apply in order to single out the storms causing erosion and to divide rainfalls of large duration in RUSLE2 [10, 11] are:

1. A rainfall event is divided into two parts, if its cumulative depth for duration of 6 hours at a certain location is less than 1.27 mm.

2. A rainfall is considered erosive if it has a cumulative value greater than 12.7 mm.

3. All rainfalls with extreme EI_{30} values and a return period greater than 50 years are deleted.

The current revised version of USLE, RUSLE2 [10], introduced the erosivity density (ED), as a measure of rainfall erosivity per unit rainfall to develop erosivity values for the USA, because ED requires shorter record lengths, as 10 years leads to acceptable results, allows more missing data than R, and is independent of the elevation:

$$ED_j = \frac{\sum_{k=1}^{m_j} (EI_{30})_k}{P_j} \tag{7}$$

where m_j is the number of storms during a time period j, $(EI_{30})_k$ the erosivity of storm k, and P_j the total precipitation height for the period j (usually monthly or annual).

Vantas et al. [12] using a numerical scheme with data from Greece showed that ED is more robust against the presence of missing precipitation values, as reflected in the following results:

- Using only 5% of the data, annual R values are underestimated on average by 85%, when the average estimation error of ED values is 50%.

- In the presence of 50% of the data, R values are underestimated by 49%, while at the same time, the estimation error of ED is 20%.

The use of ED permits the utilization of data consisting of daily rainfall values that are more abundant in comparison to data from pluviographs. The details of the method, which was employed in the USA in combination with kriging, may be found in the RUSLE2 Science Documentation [10].

3. Methods for the estimation of *R*

The lack of dense time series from pluviographs gives rise to many difficulties in the application of USLE and of its revisions in many countries. In order to cope with these difficulties, many models have been developed, based on rainfall depth values for various time steps (daily, monthly, and yearly), under specific spatial parameters and climatological data. Thus, a number of researchers reported good correlation results between *R* and yearly rainfall in many countries of the world by using various schemes ranging from simple parametric equations to geostatistical models.

These methods were applied in the countries of West Africa [13]; Switzerland taking account of elevation, aspect, longitude, and altitude [14]; the USA [15]; India [16]; Spain [17]; Italy [18]; China taking account of maximal rainfalls of 1-hour and 1-day durations per year [19]; Korea [20]; Mexico [21]; Honduras in combination with elevation [22]; Rhodesia [23]; and Hawaii [24] as well as for the development of a European [25] and Universal model [26].

A different model followed in several countries concerns the correlation, by the use of parametric equations, of the yearly values of *R* to the monthly values of rainfall depth. Those equations were applied in Venezuela [27], Germany [28], the USA [15], Italy [29], Iran in combination with the yearly and the maximum daily rainfall per year [30], North Africa [31], Morocco [32], Nigeria [33], South Italy and Southeast Australia [34], Uruguay [35], and Sudan [36].

Finally, for the estimation of the yearly value of *R*, several researchers used parametric equations for the initial determination of the daily *R* value and the corresponding daily rainfall depths. Those equations were applied in the Eastern USA [37], Australia [38], Spain [17], Kenya [39], China [40], Nigeria [41], North Africa [42], South Italy [43], Peru [44], and Slovenia [45].

Indicatively, some of the empirical equations of the literature are given below in dimensionless form, representing various parametrizations and methodologies. In West African countries, Roose [13] developed a simple linear relation between yearly values of *R* and rainfall *P*:

$$R = \alpha_1 + \beta_1 \cdot P \tag{8}$$

where α_1 και β_1 are linear regression parameters.

In the USA, Renard and Freimund [15] proposed the following nonlinear equations for continental regions of the country, in which no rainfall intensity data exist, also for yearly values of *R* and rainfall *P*:

$$R = \alpha_2 \cdot P^{\beta_2} \tag{9}$$

or

$$R = \alpha_3 + \beta_3 \cdot P^{\gamma_3} \tag{10}$$

where α_2, β_2, α_3, β_3, and γ_3 are nonlinear regression parameters.

In Morocco, Arnoldus [32] used the modified Fournier index, which is equal to

$$\mathrm{MFI} = \frac{\sum_{i=1}^{12} p_{m,i}}{P} \tag{11}$$

where *P* is the yearly rainfall and $p_{m,i}$ the monthly rainfall from January to December, and developed the formula

$$R = \alpha_5 \cdot MFI^{\beta_5} \tag{12}$$

where α_5 and β_5 are nonlinear regression parameters.

Richardson et al. [37] used the daily rainfall values for the estimation of the corresponding daily erosivity values R_d:

$$R_d = \alpha_{m,s} \cdot P_d^{\beta_{m,s}} \tag{13}$$

where the parameters $\alpha_{m,s}$ and $\beta_{m,s}$ correspond to station s and to month m and are evaluated by means of nonlinear regression.

A serious issue with the applicability of Eq. (13) is the difficulty to compute its coefficients in dry areas or months due to small number of erosive events. Another issue, even in wet periods, is the amount of missing values that also leads to a small set of observations and the same problems [46].

In order to reduce the computation burden of Eq. (13), Yu and Rosewell [38] proposed the following empirical equation:

$$R_d = \alpha_s \cdot \left[1 + \beta_s \cdot \cos\left(\frac{\pi}{6}m - \omega\right)\right] P_d^{\gamma_s} \tag{14}$$

where the parameters α_s, β_s, and γ_s are different only for the different stations s and ω is the month with the highest average of daily R values.

It must be noted that the logarithmic transformation and the subsequent application of linear regression must be avoided in the above equations, because in that case the minimization is applied to the average of the differences of the logarithms of the aggregated daily EI_{30}values (coming from pluviograph data) and the logarithms of the estimated R_d values (coming from daily rainfall data) and not of the corresponding absolute values themselves. Specifically, the application of the log transformation to Eq. (13) resulted in a −10% systematic error, when data from RUSLE2 was used, as noted on p. 59 of [10].

A different approach for the estimation of R, in the presence of missing values, was followed by Vantas and Sidiropoulos [47]. They compared empirical equations to machine learning methods, extracting for the latter methods better results, in terms of statistical significance, compared to the former methods.

4. Utilization of neural networks for the estimation of *R*

4.1 Data

The data utilized in the analysis were taken from the Greek National Bank of Hydrological and Meteorological Information [48] and came from 84 meteorological stations (**Figure 2**). By adding up the years of record registered on the various stations, the time series comprised a total of 2425 years of pluviograph records, with an average of 28.9 years per station. The time step was 30 min for the time period from 1953 to 1997, and the data coverage was equal to 56%.

For the above described data, it was deemed useful to change the time step and aggregate the data into weekly values, because 57% of the recordings were associated with storms occupying time periods covering parts of more than one calendar day, although only 17% of the storms had duration of more than 24 hours.

Under the time step of 1 week, it was found out that 80% of the values emanated from a single storm. The storms that were crossed temporally by 2 consecutive weeks were assigned to the first of the 2 weeks, and they comprised only 7% of the data. Thus, through the use of weekly instead of daily values, divisions of storms due to time step were prevented, and a measure of rainfall duration in days has been added to the observations.

4.2 Exploratory data analysis

Table 1 gives the concise statistics of the cumulative rainfall, the duration, and the erosivity EI_{30} of the erosive events that resulted following RUSLE2 rules.

In the sequel, a series of three figures is given that contain:

1. The relation of the precipitation height of the erosive events to the EI_{30} value in **Figure 3**, where a linear relation appears with a wide opening between the logarithms of rainfall values and the logarithms of EI_{30}.

2. The relation of EI_{30} to the months in **Figure 4**, where it turns out that in July its maximum appears with an almost sine-like variation of the median values per month.

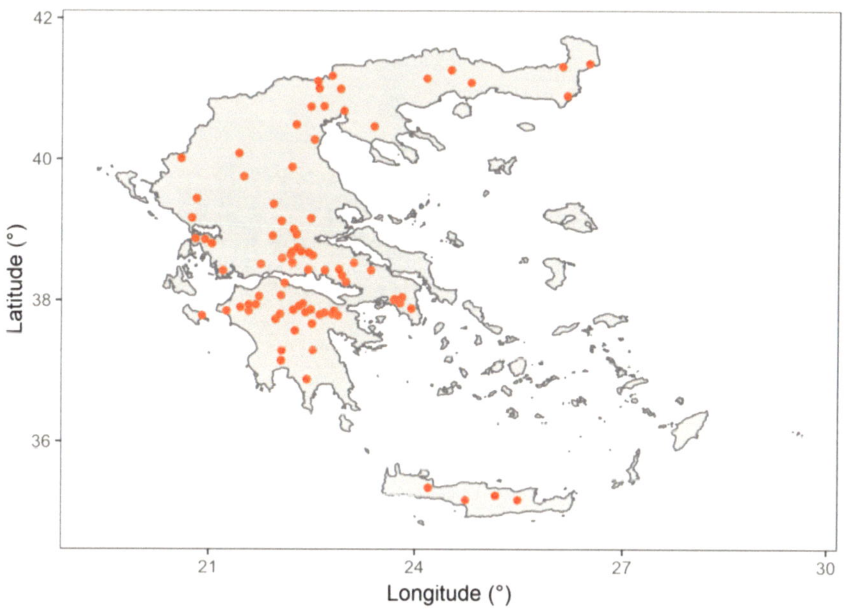

Figure 2.
Stations' location coming from Greece used in the analysis.

Variable	Min	Mean	Median	Max	SD	SW	KR	CV
EI_{30} (MJ mm ha^{-1} h^{-1})	0.7	89.1	44.2	3845.0	157.4	7.6	101.0	1.7
Precipitation (mm)	12.7	29.4	22.2	322.4	21.9	3.4	19.8	0.7
Duration (h)	0.5	13.0	10.5	152.5	10.5	2.5	13.6	0.8

SD is an abbreviation for standard deviation, SW for skew, KR for kurtosis, and CV for coefficient of variation.

Table 1.
Average statistical properties of cumulative rainfall, the duration and EI_{30} of the erosive events.

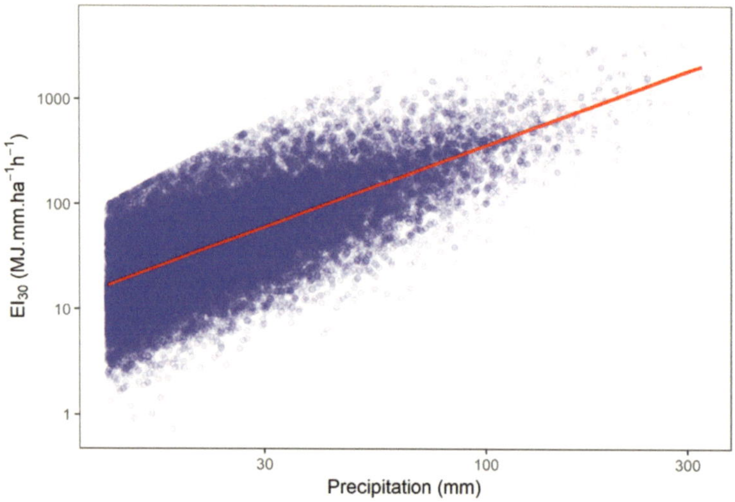

Figure 3.
Relation of cumulative precipitation of the erosive events to the EI_{30}. The red line is the LOESS line.

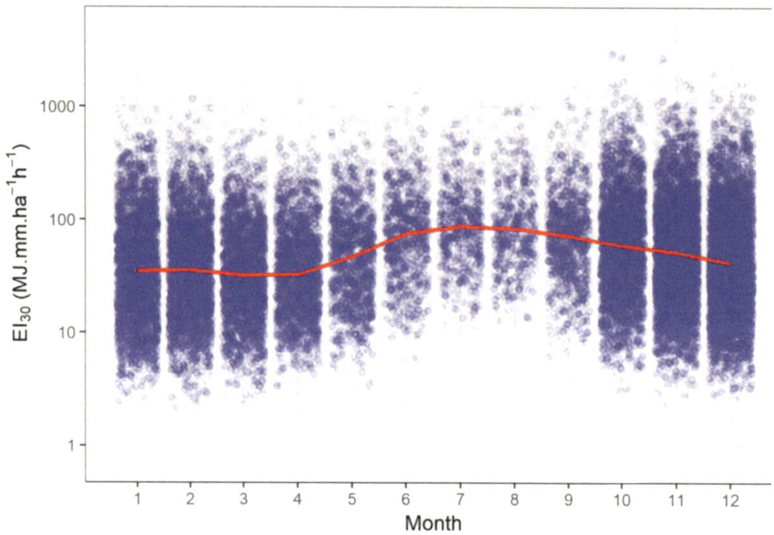

Figure 4.
Relation of cumulative precipitation of the erosive events to the EI_{30}. The red line is the LOESS line that is fitted using the median values of EI_{30} per month.

3. The relation of the duration of erosive events to EI_{30} in **Figure 5**, where it is shown that for the selected duration widths, the relation to the median values of EI_{30} is parabolic.

In these three figures, local polynomial regression fitting (LOESS, [49]) was used as a nonparametric method to produce smooth curves through the plotted variables. In that method, the fitting is done locally for a given point x using all the points in the neighborhood of x and the distances of these points from x as weights.

From the above diagrams and from the statistics of the erosive events, it can be seen that the estimation of EI_{30} values is a nonlinear problem with skewed data and with expected results of a relatively low accuracy. This point is particularly reinforced by **Figure 3**, in which it is shown that for a specific rainfall depth, such as

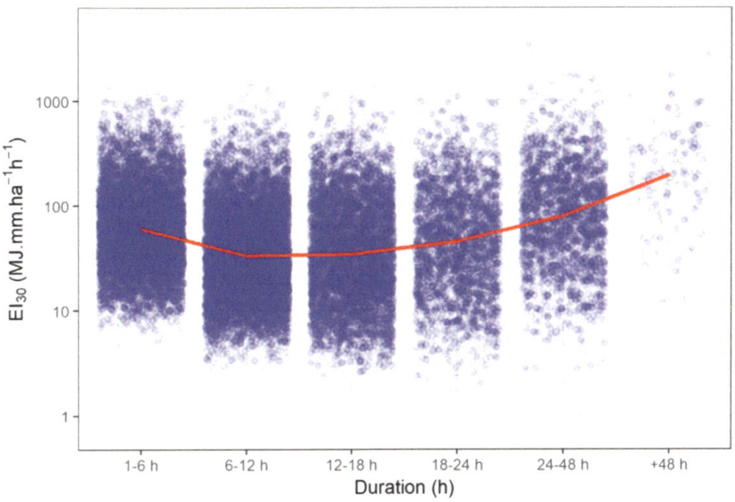

Figure 5.
Relation of cumulative precipitation of the erosive events to the EI_{30}. The red line is the LOESS line that is fitted using the median values of EI_{30} per duration width.

30 mm, a very large range of EI_{30} values, from about 10 to 700 MJ mm ha^{-1} h^{-1}, is to be assigned, corresponding to the erosive events connected to 30 mm.

4.3 Methods

The outcomes, denoted as $y^{(i)}$, where (i) is the index of a sample consisting of 19,688 values, represented weekly cumulative rainfall erosivity as calculated from pluviograph data. The features included the weekly cumulative rainfall of erosive events, the month to which the above individual m values are referred, the altitude of the meteorological stations, and the number of the days of the week for which rainfall is recorded. The matrix of the features is denoted as $x^{(i)}$, while $h(x^{(i)})$ denotes the output vector corresponding to the hypothesis in question (i.e., as computed either from the neural network or the empirical equations). The subscripts *test* and *train* mark quantities that belong to the testing or the training set, respectively.

The erosivity estimation problem is set up as a scheme of machine learning. The data were split using 70% of them as the training set and 30% as a testing one. As measures of the out-of-sample error were used:

(a) The coefficient of determination:

$$R^2 = 1 - \frac{\sum_{i=1}^{m} \left(h\left(x_{test}{}^{(i)}\right) - y_{test}{}^{(i)}\right)^2}{\sum_{i=1}^{m} \left(\hat{y}_{train} - y_{test}^{(i)}\right)^2} \tag{15}$$

with this form of R^2:

- $R^2 = 0$ means that there is no improvement comparing to a simplistic model that returns the average of the training set values $\hat{y}_{train}$.
- $R^2 = 1$ means a perfect algorithm.
- $R^2 < 0$ means an algorithm that makes prediction worse than the base model.

(b) The root mean-squared error:

$$RMSE = \sqrt{\frac{1}{m}\sum_{i=1}^{m}\left(h(x_{test}{}^{(i)}) - y_{test}{}^{(i)}\right)^2} \tag{16}$$

(c) The mean absolute error:

$$MAE = \frac{1}{m}\sum_{i=1}^{m}\left|h\left(x_{test}{}^{(i)}\right) - y_{test}{}^{(i)}\right| \tag{17}$$

In Eqs. (15)–(17), m is the number of the samples in the test set, $h\left(x_{test}{}^{(i)}\right)$ is the estimations from a trained algorithm (i.e., either from a trained neural network or from a fitted empirical equation) used as inputs, the testing set x_{test} and y_{test} are the outcomes from calculations by means of pluviograph data, and $\hat{y}_{train}$ is the average of the training set erosivity values coming also from pluviograph data.

In order to compare neural network performance to that of empirical equations, two alternative exponential models discussed in the previous section were tried, namely, those given by Eqs. (13) and (14) and leading to optimal adjustments of two respective hypotheses. The latter were determined by minimizing the following objective function by the trust-region-reflective method [50, 51]:

$$J(\theta) = \frac{1}{2m}\sum_{i=1}^{m}\left(h_\theta\left(x^{(i)}\right) - y^{(i)}\right)^2 \tag{18}$$

where θ denotes the vector of the respective parameters and $y^{(i)}$ are the outcomes, as defined in the beginning of this section.

Neural networks aim at mimicking the function of the brain. They constitute powerful nonlinear regression methods. The result or exit of the neural network is produced via a series of intermediate nodes arranged in successive layers, characterized as hidden. These layers mediate between an initial layer of input nodes and a final layer of exit nodes. The hidden nodes define linear combinations of the data contained in the initial input layer. These linear combinations are further transformed by a nonlinear function, which possesses a continuous derivative, such as the hyperbolic tangent function:

$$g(x) = \tanh(x) \tag{19}$$

The example in **Figure 6** displays three layers, of which the one to the left is the input layer, the middle one is the hidden layer, and the third one is the output layer. In the neural network, $\alpha_i^{(j)}$ denotes the activation of node i in layer j, and $\Theta^{(j)}$ denotes the matrix of weights that serve as coefficients of the linear transformation from layer j to layer $j+1$. The values at the output nodes in the example of **Figure 6** are:

$$\begin{aligned}
\alpha_1^{(2)} &= g\left(\Theta_{10}^{(1)}\cdot x_0 + \Theta_{11}^{(1)}\cdot x_1 + \Theta_{12}^{(1)}\cdot x_2 + \Theta_{13}^{(1)}\cdot x_3\right)\\
\alpha_2^{(2)} &= g\left(\Theta_{20}^{(1)}\cdot x_0 + \Theta_{21}^{(1)}\cdot x_1 + \Theta_{22}^{(1)}\cdot x_2 + \Theta_{23}^{(1)}\cdot x_3\right)\\
\alpha_3^{(2)} &= g\left(\Theta_{30}^{(1)}\cdot x_0 + \Theta_{31}^{(1)}\cdot x_1 + \Theta_{32}^{(1)}\cdot x_2 + \Theta_{33}^{(1)}\cdot x_3\right)\\
\alpha_4^{(2)} &= g\left(\Theta_{40}^{(1)}\cdot x_0 + \Theta_{41}^{(1)}\cdot x_1 + \Theta_{42}^{(1)}\cdot x_2 + \Theta_{43}^{(1)}\cdot x_3\right)\\
\alpha_1^{(3)} &= g\left(\Theta_{10}^{(2)}\cdot \alpha_0^{(2)} + \Theta_{11}^{(2)}\cdot \alpha_1^{(2)} + \Theta_{12}^{(2)}\cdot \alpha_2^{(2)} + \Theta_{13}^{(2)}\cdot \alpha_3^{(2)} + \Theta_{14}^{(2)}\cdot \alpha_4^{(2)}\right)
\end{aligned} \tag{20}$$

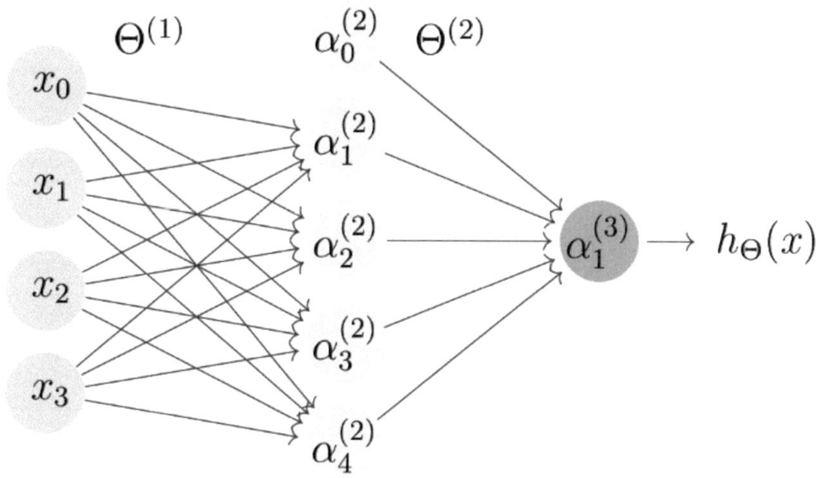

Figure 6.
A neural network with a single hidden layer.

The optimization of the $\Theta^{(j)}$ weights is executed in such a way as to minimize the sum of squares of the differences between computed and measured results:

$$J(\Theta) = \frac{1}{2m} \sum_{i=1}^{m} \left(h_\Theta\left(x^{(i)}\right) - y^{(i)}\right)^2 \tag{21}$$

where m is the number of values $h_\Theta\left(x^{(i)}\right)$ obtained from the network using $x^{(i)}$ as input data and $y^{(i)}$ as the calculated values (i.e., the EI_{30} values as computed from the pluviographic data).

This minimization is a difficult optimization problem, since there are no constraints related to these parameters. The parameters are usually initialized by random values, and in the sequel, specialized algorithms are used for their determination. A survey about neural networks and their forms as shallow and deep can be found in Schmidhuber [52] and Goodfellow et al. [53].

The architecture of the neural network used in the analysis included two hidden layers that had 32 and 16 neurons, respectively. In order to keep $h_\Theta\left(x^{(i)}\right)$ nonnegative, the second hidden layer of the neural network used the rectifier activation function [54]:

$$g(x) = \max(0, x) \tag{22}$$

The training set values x_{train} were used to compute the averages $\overline{x}$ and standard deviation $sd(x)$ and normalize both x_{train} and x_{test} using the normalizing transformation:

$$N\left(x^{(i)}\right) = \frac{x^{(i)} - \overline{x}}{sd(x)} \tag{23}$$

The training of the networks was performed by the method of early stopping [55] by utilizing a random validation set consisting of 10% of the training data, so as to avoid over fitting of the neural network.

The data importing, calculation of EI_{30} values, and analysis were done using the R language for statistical computing and graphics [56] using the packages:

Method	RMSE	MAE	R^2
Richardson et al.	141.94	65.46	0.59
Yu and Rosewell	140.36	64.33	0.60
Neural network	116.83	54.33	0.73

R^2 *values are unitless and RMSE and MAE values are in MJ mm ha*$^{-1}$ *h*$^{-1}$*. The metrics refer to the weekly erosivity time step.*

Table 2.
Estimation of the out-of-sample error metrics for the two empirical equations and the neural network.

Method	RMSE	MAE	R^2
Richardson et al.	261.73	144.63	0.76
Yu and Rosewell	259.80	141.48	0.76
Neural network	223.01	124.24	0.82

R^2 *values are unitless and RMSE and MAE values are in MJ mm ha*$^{-1}$ *h*$^{-1}$*y*$^{-1}$*. The values are calculated using the estimation of the annual erosivity values.*

Table 3.
Estimation of the out-of-sample error metrics for the two empirical equations and the neural network.

hydroscoper [57], hyetor [58], and ggplot2 [59]. The open source software library TensorFlow [60] was used in order to train the neural network.

4.4 Results

The comparison of the algorithms was based on their performance on the testing set using the three different error metrics and two time steps: weekly and annual.

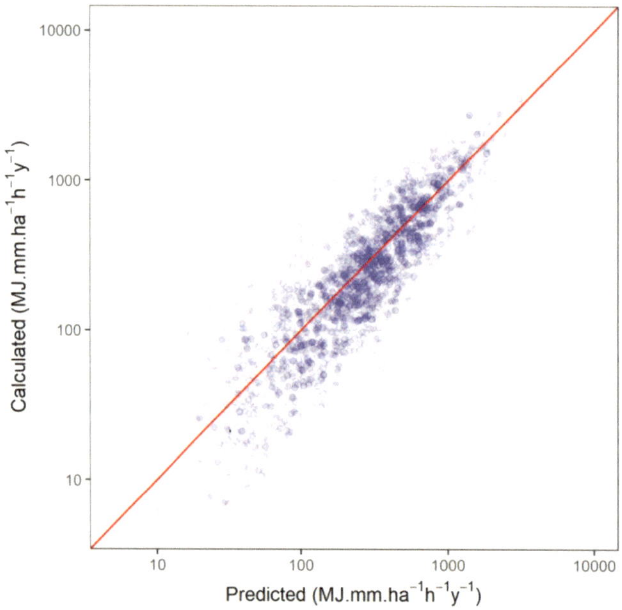

Figure 7.
Calculated annual erosivity values coming from pluviograph data versus predicted values using the Yu and Rosewell model on the testing set. With red line, the identity function $f(x) = x$ *is symbolized.*

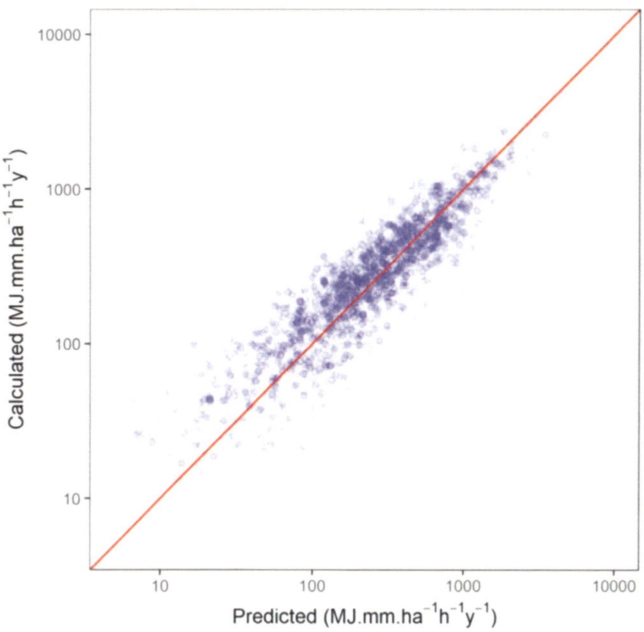

Figure 8.
Calculated annual erosivity values coming from pluviograph data versus predicted values using the neural network on the testing set. With red line, the identity function $f(x) = x$ is symbolized.

The weekly and the annual calculated values were produced from the aggregated erosivity values coming from the pluviograph data. The annual predicted values are also the aggregated values coming from the output of the corresponding algorithm (the predicted weekly erosivity values in our case). The annual time step was used in order to examine if there is a systematic error on the estimations of the algorithms, which was not the case.

These results are given in **Tables 2** and **3**, where it is shown that the neural network outperformed the two parametric methods, which both had similar results. Specifically, the machine learning method had better performance that ranged from 8 to 22%, depending on the error metric and time step.

In **Figures 7** and **8**, the annual erosivity values are presented as calculated in the test set versus the values estimated from the 84 fitted Yu and Rosewell equations for each station and the single neural network. In these two diagrams, the neural network's estimations are closer to the calculated ones than those coming from the empirical equations.

5. Conclusions

The universal soil loss equation with its revisions represents the established model and the means for estimating annual long-term average soil loss. An important factor of this equation is the rainfall erosivity, R, which is closely connected to the energy carried by rainfall. Rainfall data are essential for the estimation of R. In particular, data from pluviographs are more suitable for that task. However, such data are scarce, and various empirical methods have been presented for estimating R from coarser data. These methods are reviewed in this chapter.

Also, various difficulties are discussed that are associated with the issues of missing values and the separation of erosive events. In addition, indications are given of the adverse nature of the relation of erosivity to precipitation characteristics. All these facts raise objections to the suitability of empirical models and lead to the use of data-driven methods that will be able to handle more effectively the problematic nature of the data and also to discover more reliably the hidden relations between erosivity, precipitation, seasonality, and local characteristics. This is achieved, because in machine learning procedures, no human intervention or bias is involved in the selection or in the forms of these hidden nonlinearities.

More specifically, neural networks, as machine learning tools, are employed in this chapter for the estimation of erosivity, in comparison to two different empirical equations of the literature. The results demonstrate, first of all, the superior performance of the machine learning method and its ability for generalization, so as to perform equally well upon new data. Another advantage of the proposed learning algorithm is that it produces a single model accessing all available data, in contrast to empirical equations, where a researcher must fit and handle as many nonlinear equations as the number of available stations or the product of the number of the stations by the number of the used seasons, making the analysis cumbersome and prone to errors.

The latter fact is exemplified on the country-wide data from Greece, used in this chapter. The single neural network, which was produced, outperformed 81 different empirical nonlinear equations, each one of which was fitted using data from each station. The comparison of the performance was made using test data sets to which neither the empirical equations nor the neural network had any access during their training.

In conclusion, the present chapter gives an indicative and characteristic sample of the use of machine learning methods in the problems associated with erosivity. There are still many related problems, such as the spatial and temporal distribution of erosivity, with significant potential for the application of machine learning methods. The same is true of many areas of hydrology and water resources.

Conflict of interest

The authors declare no conflict of interest.

Author details

Konstantinos Vantas, Epaminondas Sidiropoulos* and Chris Evangelides
Faculty of Engineering, Aristotle University of Thessaloniki, Thessaloniki, Macedonia, Greece

*Address all correspondence to: nontas@topo.auth.gr

References

[1] Boardman J, Poesen J. Soil erosion in Europe: Major processes, causes and consequences. In: Soil Erosion in Europe. Wiley-Blackwell; 2006. pp. 477-487

[2] Kinnell PIA. Event soil loss, runoff and the universal soil loss equation family of models: A review. Journal of Hydrology. 2010;**385**(1):384-397

[3] Nearing MA, Yin S, Borrelli P, Polyakov VO. Rainfall erosivity: An historical review. Catena. 2017;**157**: 357-362

[4] Wischmeier WH, Smith DD. Predicting Rainfall Erosion Losses—A Guide to Conservation Planning. USDA, Science and Education Administration; 1986

[5] Renard KG, Foster GR, Weesies GA, McCool DK, Yoder DC. Predicting Soil Erosion by Water: A Guide to Conservation Planning with the Revised Universal Soil Loss Equation (RUSLE). Vol. Agriculture Handbook No. 703. Washington, DC: United States Department of Agriculture; 1997

[6] Brown L, Foster G. Storm erosivity using idealized intensity distributions. Transactions of the ASAE. 1987; **30**(2):379-386

[7] Ramon R, Minella JP, Merten GH, de Barros CA, Canale T. Kinetic energy estimation by rainfall intensity and its usefulness in predicting hydrosedimentological variables in a small rural catchment in southern Brazil. Catena. 2017;**148**:176-184

[8] McGregor KC, Mutchler CK. Status of the R factor in northern Mississippi. In: Soil Erosion: Prediction and Control. Soil Conservation Society of America; 1976. pp. 135-142

[9] Van Dijk A, Bruijnzeel LA, Rosewell CJ. Rainfall intensity–kinetic energy relationships: A critical literature appraisal. Journal of Hydrology. 2002; **261**(1–4):1-23

[10] USDA-ARS. Science Documentation, Revised Universal Soil Loss Equation Version 2 (RUSLE2). USDA-Agricultural Research Service; 2013

[11] Yin S, Nearing MA, Borrelli P, Xue X. Rainfall Erosivity: An overview of methodologies and applications. Vadose Zone Journal. 2017;**16**(12):0

[12] Vantas K, Sidiropoulos E, Loukas A. Temporal and elevation trend detection of rainfall erosivity density in Greece. In: 3rd International Electronic Conference on Water Sciences (ECWS-3); 2018

[13] Roose E. Erosion et ruissellement en Afrique de l'Ouest, vingt années de mesures en parcelles expérimentales. O. R.S.T.R.O.M. 1977;**78**:30-33

[14] Meusburger K, Steel A, Panagos P, Montanarella L, Alewell C. Spatial and temporal variability of rainfall erosivity factor for Switzerland. Hydrology and Earth System Sciences. 2012; **16**(1):167-177

[15] Renard KG, Freimund JR. Using monthly precipitation data to estimate the R-factor in the revised USLE. Journal of Hydrology. 1994;**157**:287-306

[16] Sudhishri S, Patnaik U. Erosion index analysis for Eastern Ghat high zone of Orissa. Indian Journal of Dryland Agricultural Research and Development. 2004;**19**:42-47

[17] Angulo-Martínez M, Beguería S. Estimating rainfall erosivity from daily precipitation records: A comparison among methods using data from the

Ebro Basin (NE Spain). Journal of Hydrology. 2009;**379**(1):111-121

[18] Diodato N. Estimating RUSLE's rainfall factor in the part of Italy with a Mediterranean rainfall regime. Hydrology and Earth System Sciences Discussions. 2004;**8**(1):103-107

[19] Zhang G, Nearing M, Liu B. Potential effects of climate change on rainfall erosivity in the Yellow River Basin of China. Transactions of the ASAE. 2005;**48**(2):511-517

[20] Lee J-H, Heo J-H. Evaluation of estimation methods for rainfall erosivity based on annual precipitation in Korea. Journal of Hydrology. 2011;**409**(1):30-48

[21] Millward AA, Mersey JE. Adapting the RUSLE to model soil erosion potential in a mountainous tropical watershed. Catena. 1999;**38**(2):109-129

[22] Mikhailova EA, Bryant R, Schwager S, Smith S. Predicting rainfall erosivity in Honduras. Soil Science Society of America Journal. 1997;**61**(1):273-279

[23] Stocking M, Elwell H. Rainfall erosivity over Rhodesia. Transactions of the Institute of British Geographers. 1976;**2**:231-245

[24] Lo A, El-Swaify S, Dangler E, Shinshiro L. Effectiveness of EI30 as an erosivity index in Hawaii. Soil Conservation Society of America. 1985;**1**:384-339

[25] Panagos P, Ballabio C, Borrelli P, Meusburger K, Klik A, Rousseva S, et al. Rainfall erosivity in Europe. Science of the Total Environment. 2015;**511**: 801-814

[26] Yang D, Kanae S, Oki T, Koike T, Musiake K. Global potential soil erosion with reference to land use and climate changes. Hydrological Processes. 2003; **17**(14):2913-2928

[27] Lujan DL, Gabriels D. Assessing the rain erosivity and rain distribution in different agro-climatological zones in Venezuela. Sociedade & Natureza. 2005;**1**(1):16-29

[28] Sauerborn P, Klein A, Botschek J, Skowronek A. Future rainfall erosivity derived from large-scale climate models—Methods and scenarios for a humid region. Geoderma. 1999;**93**(3): 269-276

[29] Lastoria B, Miserocchi F, Lanciani A, Monacelli G. An estimated erosion map for the Aterno-Pescara River basin. European Water. 2008;**21**(22):29-39

[30] Sepaskhah A, Sarkhosh P. Estimating storm erosion index in southern region of Iran. Iranian Journal of Science and Technology, Transaction B: Engineering. 2005;**29**(3):357-363

[31] Smithen A, Schulze R. The spatial distribution in Southern Africa of rainfall erosivity for use in the universal soil loss equation. Water SA. 1982;**8**(2): 74-78

[32] Arnoldus H. Methodology used to determine the maximum potential average annual soil loss due to sheet and rill erosion in Morocco. FAO Soils Bulletins (FAO). 1977;**34**:8-9

[33] Igwe C, Mbagwu J, Akamigbo F. Application of SLEMSA and USLE models for potential erosion hazard mapping in south-eastern Nigeria. International Agrophysics. 1999;**13**:41-48

[34] Ferro V, Porto P, Yu B. A comparative study of rainfall erosivity estimation for southern Italy and Southeastern Australia. Hydrological Sciences Journal. 1999;**44**(1):3-24

[35] Munka C, Cruz G, Caffera RM. Long term variation in rainfall erosivity in Uruguay: A preliminary Fournier approach. GeoJournal. 2007;**70**(4): 257-262

[36] Elagib N. Changing rainfall, seasonality and erosivity in the hyper-arid zone of Sudan. Land Degradation & Development. 2011;**22**(6):505-512

[37] Richardson C, Foster G, Wright D. Estimation of erosion index from daily rainfall amount. Transactions of the ASAE. 1983;**26**(1):153-156

[38] Yu B, Rosewell C. An assessment of a daily rainfall erosivity model for New South Wales. Soil Research. 1996;**34**(1): 139-152

[39] Angima S, Stott D, O'Neill M, Ong C, Weesies G. Soil erosion prediction using RUSLE for central Kenyan highland conditions. Agriculture, Ecosystems & Environment. 2003; **97**(1):295-308

[40] Xie Y, Yin S, Liu B, Nearing MA, Zhao Y. Models for estimating daily rainfall erosivity in China. Journal of Hydrology. 2016;**535**:547-558

[41] Salako F. Development of isoerodent maps for Nigeria from daily rainfall amount. Geoderma. 2010; **156**(3):372-378

[42] Le Roux J, Morgenthal T, Malherbe J, Pretorius D, Sumner P. Water erosion prediction at a national scale for South Africa. Water SA. 2008;**34**(3):305-314

[43] Capolongo D, Diodato N, Mannaerts CM, Piccarreta M, Strobl R. Analyzing temporal changes in climate erosivity using a simplified rainfall erosivity model in Basilicata (Southern Italy). Journal of Hydrology. 2008; **356**(1):119-130

[44] Elsenbeer H, Cassel D, Tinner W. A daily rainfall erosivity model for Western Amazonia. Journal of Soil and Water Conservation. 1993;**48**(5): 439-444

[45] Petkovšek G, Mikoš M. Estimating the R factor from daily rainfall data in the sub-Mediterranean climate of Southwest Slovenia. Hydrological Sciences Journal. 2004;**49**(5)

[46] Hollinger SE, Angel JR, Palecki MA. Spatial Distribution, Variation, and Trends in Storm Precipitation Characteristics Associated with Soil Erosion in the United States. Prepared for the United State Department of Agriculture; 2002. p. 103

[47] Vantas K, Sidiropoulos E. Imputation of erosivity values under incomplete rainfall data by machine learning methods. European Water. 2017;**57**:193-197

[48] Vafiadis M, Tolikas D, Koutsoyiannis D. HYDROSCOPE: The new Greek national database system for meteorological, hydrological and hydrogeological information. In: WIT Transactions on Ecology and the Environment. Vol.3. WIT Press; 1994. pp. 1-8

[49] Cleveland WS, Grosse E, Shyu WM. Local regression models. Statistical models in S. 1992;**2**:309-376

[50] Coleman T, Li Y. On the Convergence of Reflective Newton Methods for Large-Scale Nonlinear Minimization Subject to Bounds. Vol. 67. Ithaca, NY, USA: Cornell University; 1994

[51] Coleman TF, Li Y. An interior trust region approach for nonlinear minimization subject to bounds. SIAM Journal on Optimization. 1996;**6**(2): 418-445

[52] Schmidhuber J. Deep learning in neural networks: An overview. Neural Networks. 2015;**61**:85-117

[53] Goodfellow I, Bengio Y, Courville A, Bengio Y. Deep Learning. Vol. 1. Cambridge: MIT Press; 2016

[54] Glorot X, Bordes A, Bengio Y. Deep sparse rectifier neural networks. In:

Proceedings of the Fourteenth International Conference on Artificial Intelligence and Statistics; FL, USA: Fort Lauderdale; 2011. pp. 315-323

[55] Wang C, Venkatesh SS, Judd JS. Optimal stopping and effective machine complexity in learning. Advances in Neural Information Processing Systems. 1994;**6**:303-310

[56] R Core Team. R: A Language and Environment for Statistical Computing. Vienna, Austria: R Foundation for Statistical Computing; 2018

[57] Vantas K. Hydroscoper: R interface to the Greek National Data Bank for hydrological and meteorological information. Journal of Open Source Software. 2018;**3**(23):625

[58] Vantas K. hyetor: R package to analyze fixed interval precipitation time series. 2018. Available from: https://kvantas.github.io/hyetor/

[59] Wickham H. ggplot2: Elegant Graphics for Data Analysis. New York: Springer-Verlag; 2009. p. 213

[60] Abadi M, Barham K, Chen J, Chen Z, Davis A, Dean J, et al. Tensorflow: A system for large-scale machine learning. In: OSDI. 2016. pp. 265-283

[61] Haan CT, Barfield BJ, Hayes JC. Design Hydrology and Sedimentology for Small Catchments. Academic Press; 1990. pp. 238-246

Chapter 3

Simulation of Surface Runoff and Channel Flows Using a 2D Numerical Model

Yafei Jia, Tahmina Shirmeen, Martin A. Locke, Richard E. Lizotte Jr. and F. Douglas Shields Jr.

Abstract

Numerical simulation of surface runoff is used to understand and predict watershed sediment transport and water quality and improve management of agricultural watersheds. However, models currently available are either simplified or parameterized for efficiency. In this chapter, CCHE2D, a physically based hydrodynamic model for general free surface flow hydrodynamics, was applied to study watershed surface runoff and channel flows. Multiple analytical solutions and experimental data were used to verify and validate this finite element model systematically with good results. A numerical scheme for correcting the bilinear interpolation of the water surface elevation solutions from the cell centers to the computational nodes was developed to improve the model. The correction was found necessary and effective for the sheet runoff simulations over the irregular bed topography. The modified numerical model was then used to simulate storms in a low-relief agricultural watershed in the Mississippi River alluvial plain. This physically based model identified the channel networks, watershed boundary automatically, and helped to develop rating curves at the gage station of this complex watershed. The numerical simulations resolved detailed runoff and turbulent channel flows, which can be used for soil erosion and gully development analyses.

Keywords: overland flow, rainfall runoff, numerical modeling, physically based model, model validation, soil erosion

1. Introduction

Numerical simulation is increasingly used for studying overland flows. Since runoff drives soil erosion and landscape evolution, the runoff models provide a foundation for modeling soil erosion, rill erosion, and related processes at the watershed scale [1, 2]. Models involving different levels of abstraction have been proposed [3–5]. Two commonly used models are the diffusion wave (DW) and kinematic wave (KW) models [6–9]. The KW models set the friction slope to be equal to the bed slope and ignore the inertial terms [10]. The method has been successfully used to describe overland flows [11–14]. The governing equations are highly nonlinear and do not have general analytical solutions, so one has to solve them numerically for practical cases [15]. The models based on full Saint-Venant (SV) equations have also been applied and produced better results.

Two-dimension models are generally used for cases with irregular domains. A distributed rainfall-runoff model using the KW approximation solved by an implicit finite difference scheme was developed [16], but channel flows are computed using a separate KW model. Fully two-dimensional shallow water equations are being utilized for modeling overland flows in late 1980s [17]. A two-dimensional finite difference (FD) runoff model was developed by solving 2D SV equations [18]. Shallow water equation-based 2D models [19] were used for runoff over an irregular topography of experimental scale with infiltration processes considered and in rural semiarid watersheds for overland flows generated by storms [20].

In addition to finite difference method (FDM), the two-dimensional finite element (FEM) and finite volume methods (FVM) have been used for overland flow simulations. A FEM KW model was developed by Liu et al. [21] for simulating runoff generation and concentration over an irregular bed and reproduced experimental results. Tests [15] indicated that the FVM-based 2D SV model performed better than that of FDM. Costabile et al. [22] solved the shallow water equations using the FVM and applied the resulting model to simulate a real event on a watershed of 40 km^2. Nunoz-Carpena et al. [23] solved the KW equation using the Petrov-Galerkin method. Venkata et al. [24] developed a Galerkin DW FEM and applied it to a small watershed. Singh et al. [25] simulated runoff processes by solving the 2D shallow water equations with a shock-capturing scheme and the FVM. Shirmeen et al. [26] showed results of a validated, FEM 2D model in predicting runoff from a flat agricultural watershed.

In order to check numerical models' mathematical correctness and physical applicability, the developed computational models have been tested with analytical solutions, experimental, and field data. Iwagaki [27] studied runoff using analytical methods and experimental data; several specific solutions were developed based on the characteristic method. Govindaraju et al. [28] developed analytical solutions using KW and DW approximations. Comparisons of analytical solutions, numerical solutions, and experimental data were discussed. Singh [29] detailed the KW model's analytical and numerical solutions and their wide applications. Cea [30] tested FVM using an experimental watershed with a complex shape. These overland flow models use simplified equations and need to specify pre-existing channel networks, which make it difficult to simulate soil erosion cases with hill-slope evaluation and mixed sheet-channel flow conditions.

CCHE2D is a physically based model, which treats the entire watershed including the channels and ditches as one continuous domain. One does not need to differentiate overland sheet flow and channel flow calculation areas using grid cells and 1D channel networks as is done in GSSHA [31], WASH123D [32], NIKE-SHE [33], and SHETRAN [34]. It is also not necessary to employ arbitrarily shaped sub-watersheds and 1D channel networks as is done in the CCHE1D model [35]. In these models, 2D DW equations or KW equations are solved for the overland flow using finite difference methods, and the 1D SV equation is solved in the prescribed channel networks. In contrast to these models, in CCHE2D, hydrodynamics over the entire watershed is simulated using only 2D equations discretized on an irregular quadrilateral finite element mesh, which is generated using digital elevation model (DEM) data. The simulated overland sheet flow and channel flow are seamlessly connected everywhere in the domain and the channel network is formed automatically. This method may be more applicable when sediment transport, rill erosion, or gully erosion processes in watersheds are considered.

In this study, the CCHE2D model is modified and applied to simulate watershed hydrological processes. CCHE2D is a general hydrodynamic model for unsteady, turbulent free flows, sediment transport, and pollutant transport. It has been validated and applied widely to simulations of channel flow, flooding, coastal flow, bed topographic change, and chemical contamination in aquatic environments [36–40].

The major objectives of the present paper are to assess the accuracy and the effectiveness of this FEM in predicting overland runoff processes, and its applicability to practical agricultural watersheds with ditches and natural stream channels. The approach of the study followed the recommendations of [41] for quality assurance that numerical models have to be verified and validated using analytical solutions, physical experimental data, and field data. The validated numerical model was used to simulate and characterize the hydrological processes of an agricultural watershed in the Mississippi River alluvial plain where farm fields are drained and separated by ditches and stream channels. A limitation was found in the interpolation method when it is applied to the water surface elevation of the sheet runoff. A numerical scheme was developed and implemented for improving the bilinear interpolation. The present study focused on watershed surface flow processes over bare soils; interception, evapotranspiration, and infiltration were not considered.

2. Mathematical model

Surface runoff due to precipitation is typically quite shallow and can be aptly represented by the 2D shallow water equations within the CCHE2D model [36, 38]. The water surface elevation of the runoff flow, η, is calculated by the continuity equation in a Cartesian coordinate system

$$\frac{\partial \eta}{\partial t} + \frac{\partial uh}{\partial x} + \frac{\partial vh}{\partial y} = R \tag{1}$$

in which h is the local water depth, t is time; R is rainfall intensity, which may vary in time and space, and u and v are depth-averaged velocity components in x and y directions, respectively. The depth-integrated 2D momentum equations for turbulent flows are as follows:

$$\frac{\partial uh}{\partial t} + \frac{\partial uuh}{\partial x} + \frac{\partial vuh}{\partial y} = -gh\frac{\partial \eta}{\partial x} + \left(\frac{\partial h\tau_{xx}}{\partial x} + \frac{\partial h\tau_{xy}}{\partial y}\right) - \frac{\tau_{bx}}{\rho} \tag{2}$$

$$\frac{\partial vh}{\partial t} + \frac{\partial uvh}{\partial x} + \frac{\partial uvh}{\partial y} = -gh\frac{\partial \eta}{\partial y} + \left(\frac{\partial h\tau_{yx}}{\partial x} + \frac{\partial h\tau_{yy}}{\partial y}\right) - \frac{\tau_{by}}{\rho} \tag{3}$$

in which g is the gravitational acceleration, ρ is water density, τ_{xx}, τ_{xy}, τ_{yx}, and τ_{yy} are depth-averaged Reynolds stresses, and τ_{bx}, τ_{by} are bed shear stresses. In the overland runoff area, the Reynolds stress terms vanish, and Eqs. (2) and (3) become the shallow water equations. The Reynolds stress terms remain significant in the part of the domain with channel and concentrated flows. A special finite element method called the efficient element method is adopted in the model, in which a collocation approach is used to discretize the equations in a structured quadrilateral nonorthogonal mesh system. A partially staggered grid is used for solving these equations. A velocity correction method is used to couple the continuity equation and the momentum equations. More details about this model's numerical methodology and techniques can be found in earlier publications [36, 38, 42].

The full Eqs. (1)–(3) are applicable for general flow conditions. In realistic cases where runoff and channel flow conditions coexist, a general flow model is necessary. Under the sheet flow condition, the advection and turbulence stress terms in the momentum equations vanish because the dominant forcing for the overland flow is the gravity and bed shear stress. The water depth is very small, and water surface slope and bed slope become almost the same:

$$\frac{\partial \eta}{\partial x} \approx \frac{\partial b}{\partial x}, \quad \frac{\partial \eta}{\partial y} \approx \frac{\partial b}{\partial y} \tag{4}$$

in which b is the bed elevation. The general flow equations then become the KW equations. Under this condition, the flow is completely dominated by the bed slope. Shear stresses on the bed are evaluated in conjunction with the Manning equation as:

$$\tau_{bx} = \frac{1}{h^{1/3}} \rho g n^2 u U, \tag{5}$$

$$\tau_{by} = \frac{1}{h^{1/3}} \rho g n^2 v U \tag{6}$$

in which n is the Manning roughness coefficient and $U = \sqrt{u^2 + v^2}$ is the total velocity magnitude.

3. Interpolation of water surface elevation

CCHE2D uses a partially staggered method: the velocities are solved at collocation points and the pressure (water surface) is solved at cell centers [36]. A bilinear interpolation method is used to interpolate the water surface elevation solution to the collocation nodes where the momentum equations are solved. The bilinear interpolation works well for general channel flow simulations because the water depth is large in comparison with the variation of bed surface and the mesh size. When overland sheet runoff is simulated; however, the water depth is very small; it is often less than the microelevation variation of bed topography represented in an element. In this case, the interpolated water surface elevation may be lower than the bed if the bed is concave down and vice versa. This is a limitation of the interpolation method. In the concave down case, dry nodes are created artificially; in the concave up case, artificial masses of water could be erroneously created. **Figure 1** illustrates this problem in one dimension. The problem occurs whenever irregular bed topographies are encountered. A correction is therefore necessary to the interpolation over the surface runoff area.

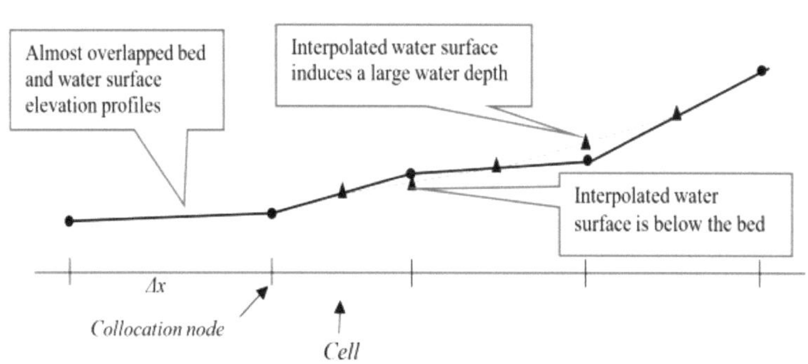

Figure 1.
The error of underestimation and overestimation caused by linear interpolation of water surface elevation from cell centers to collocation nodes.

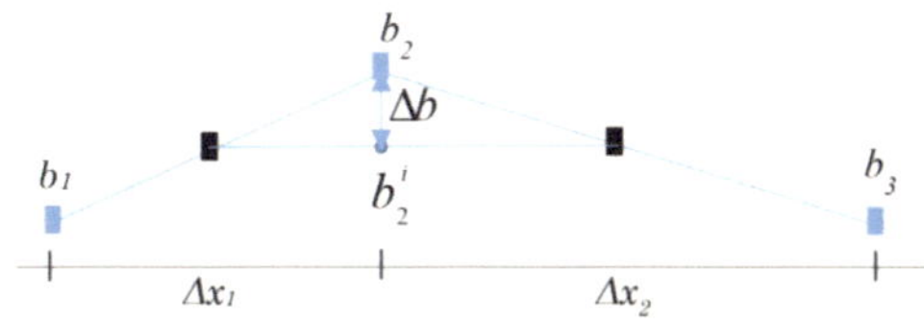

Figure 2.
Definition sketch for the formulation of the correction (Eqs. (7) and (8)) to linear water surface elevation interpolation. b_1, b_2, *and* b_3 *are bed elevation.* Δb *is the interpolation correction and* Δx_1 *and* Δx_2 *are mesh spacing.*

A numerical scheme has been developed and implemented in CCHE2D to correct the interpolation error [43]. **Figure 2** illustrates how the scheme is formulated in one dimension with an exaggerated vertical scale. Eq. (7) is the formulation to compute the correction value Δb for nonuniform meshes, and it is simplified to Eq. (8) if the mesh is uniform. It is straightforward to extend Eqs. (7) and (8) to two dimension. Water depth at the cell centers is positive, without this correction, the depth at the middle point would become negative because the interpolated water surface elevation is below the bed. This scheme is necessary and effective when cases with irregular topography are simulated

$$\Delta b = b_2 - b_2^i = \frac{1}{2}\left[b_2 - \frac{b_1 \Delta x_2}{\Delta x_1 + \Delta x_2} - \frac{b_3 \Delta x_1}{\Delta x_1 + \Delta x_2}\right] \tag{7}$$

$$\Delta b = \frac{1}{4}(b_1 - 2b_2 + b_3) \tag{8}$$

where b_1, b_2, and b_3 are bed elevation, Δb is the interpolation correction, b_2^i is the linearly interpolated value, and Δx_1 and Δx_2 are mesh spacing (**Figure 2**). The interpolated water surface elevation needs to be corrected by Δb.

4. Analytical verification

Two analytical solutions were obtained by solving a one-dimensional kinematic equation analytically for rain-generated runoff by [44, 45]. The solution of sustained rains for the runoff to reach a steady state [44] and the solution for rainfall that stops before the runoff becomes steady [45], including the tailing stage solution after rainfall stops, were provided. The governing one-dimensional kinematic equation for deriving these solutions is:

$$\frac{\partial h}{\partial t} + u\frac{\partial h}{\partial x} = R \tag{9}$$

$$u = \alpha h^{k-1}, \quad q = uh = \alpha h^k \tag{10}$$

in which q is the discharge of water per unit width (m^2/s), k is an exponent (=5/3), and α (=5) is a coefficient (m^{2-k}/s). These analytical solutions were realized for a few simple cases: runoff due to steady rainfall intensity on a uniform planar area of 200 × 1 m with a slope of 1.0%. The rainfall intensity was $R = 2.7 \times 10^{-5}$ m/s, and the Manning's coefficient was $n = 0.02\ m^{-1/3}s$. For comparison, the same case was simulated using CCHE2D and a 10 × 100 point 2D mesh with uniform spacing. The solutions were recorded at cross sections located at 50, 100, 150, and 200 m, from the upstream end of the plane.

Figure 3 shows the comparisons of the simulated runoff and the analytical solutions for the sustained rain collected at the four cross sections. Hydrographs at each cross section indicate that equilibrium runoff (steady state) is reached before the rain stops at $T = 1000$ s. The runoff is always nonuniform, and the peak discharge increases in the downstream direction. At first, the flow is unsteady (rising limb), then becomes steady until $T = 1000$ s, and finally becomes unsteady in the falling limb. The runoff reaches equilibrium earlier at locations closer to upstream. The simulation is a little less than the analytical solution at the time approaching the peak discharge, particularly near the downstream. The solution can be improved by reducing the local mesh size effectively.

Figure 4 shows a case in which the rainfall stops before runoff reaches steady state ($T = 200$ s); the hydrographs, thus, have a different pattern. The peak discharge is reached at the time the rain stopped and is the same for all cross sections. The peak discharge for the lower cross sections lasts longer because the flows at the lower locations are sustained by upstream contributions. The runoff recession is earlier for upstream locations. The shape of the two sets of simulated hydrographs at all cross-section locations corresponded well with the analytical solutions.

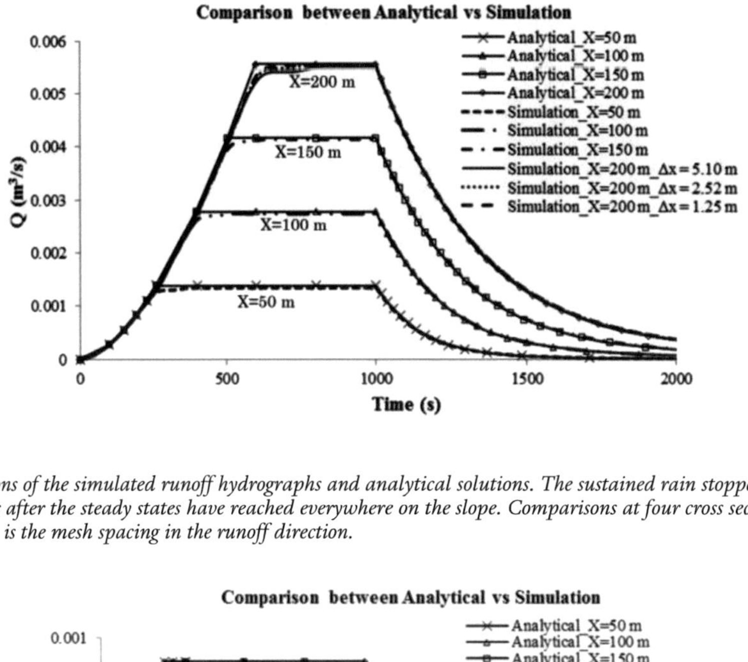

Figure 3.
Comparisons of the simulated runoff hydrographs and analytical solutions. The sustained rain stopped at T = 1000 *s after the steady states have reached everywhere on the slope. Comparisons at four cross sections are shown.* Δx *is the mesh spacing in the runoff direction.*

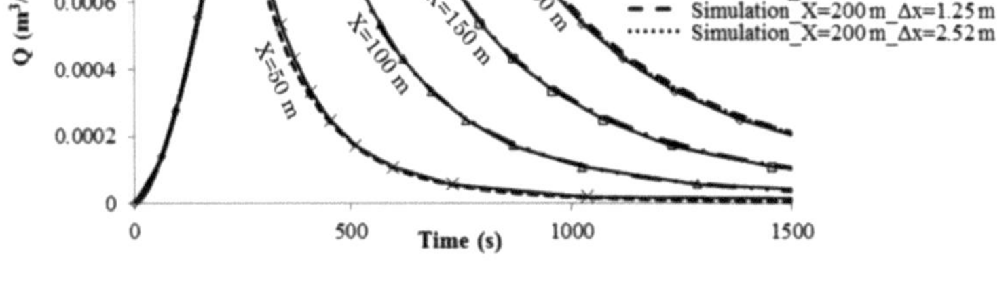

Figure 4.
Comparisons of the simulated runoff hydrographs and analytical solutions. The rain stopped at T = 200 *s, before the flow at any of the four cross sections reached steady state.* Δx *is the mesh spacing in the runoff direction.*

5. Experimental validation

CCHE2D model was validated using experimental data sets collected from the literature. All of these cases were carried out on impervious overland flow planes. The only quantity measured in these experiments was the downstream runoff discharge.

5.1 Case 1

Morgali and Linsley [46] obtained two sets of experimental runoff data. Their tests were carried out over a straight turf surface of 21.95 m long with a constant slope (0.04) and width. The Manning's coefficient, n, was found to be 0.5 $m^{-1/3}s$. The rains had two different intensities and were uniform along the slope for 1200 s (20 min). **Figure 5** compares the experimental data and the numerical simulations. The analytical solution for this test condition [44] is also presented in **Figure 5**. It was found that these runoff experiments fit well with the analytical solution. A 110 × 10 points uniform mesh and 0.01 s time step were used for the numerical simulation. The CCHE2D numerical results showed good agreements with the analytical solution as well as the experimental results (**Figure 5**). The rising limb of the discharge hydrograph and the peak discharge were captured very well by the simulations. The processes of the two experiments, 1A (R = 92.96 mm/h) and 1B (R = 48.01 mm/h), look similar because the only difference in the experiments was rainfall intensity. The peak discharges of the experiments occurred at approximately 850 and 1100 s, respectively, for Case 1A and Case 1B. The numerical solutions of CCHE2D agreed well with the experimental data. The peak discharges for Case 1A and Case 1B are 5.67×10^{-4} and 2.93×10^{-4} m^3/s, respectively. The times to peak discharge for Case 1A resulted from the analytical solution, CCHE2D

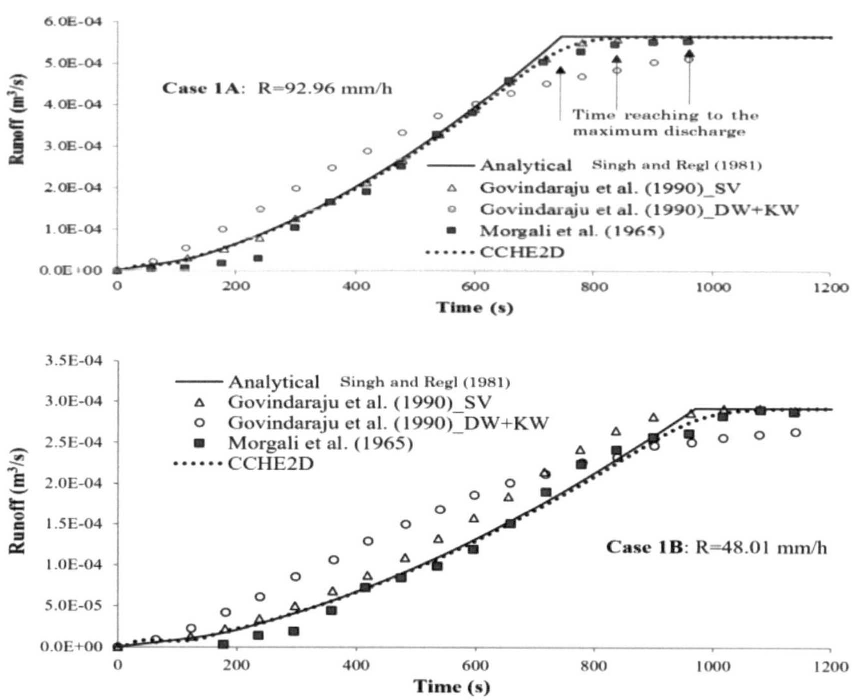

Figure 5.
Comparisons of measured data with analytical solution, results of CCHE2D, and other numerical models.

and the experiment, are 760, 850 and 950 s, respectively. The differences among the three are less for the Case 1B (**Figure 5**).

Figure 5 also compares the simulation results of CCHE2D and the model results by Govindaraju et al. [28]; the two numerical solutions agree well for the case with the higher rain (1A), but the fit of their solutions based on the SV equations does not correspond well for the case with the smaller rainfall (1B). The results of CCHE2D also outperform the analytical solution of the DW approximation [28].

5.2 Case 2

Cea et al. [30] conducted three runoff experiments of complex topography and simulated these cases using a 2D unstructured FVM. The experimental watershed was a rectangle (2 × 2.5 m) made by three planes of stainless steel, each of them with a slope of 0.05 (**Figure 6**). Two dikes (1.86 and 1.01 m in length) were placed in the watershed to vary the topography. Rainfall intensity, duration, and runoff hydrographs were measured. As a result, the runoff direction, distribution, and pattern of the hydrograph were affected. The runoff was accumulated and became channel flows along intercepting lines of slopes and dikes. Since both overland flow and channel flow are involved, faithful simulation requires solving full governing Eqs. (1)–(3). The rainfall applied to each test case was different. In the first test (2A), the rainfall intensity was 317 mm/h for 45 s. In the second test (2B), the rainfall intensity was 320 mm/h for 25 s; then it was stopped for 4 s and restarted for an additional 25 s with the same intensity. In the third test (2C), rainfall intensity was 328 mm/h. The rainfall was applied for 25 s; then it was stopped for 7 s and then restarted for another 25 s.

In this study, CCHE2D was applied and the numerical results were compared with experimental data. The watershed was modeled using an irregular structured mesh with the cell size ranging from 0.034 to 0.009 m; the mesh was refined near the main channel and the outlet for improving results. The Manning's roughness coefficient was set equal to 0.009 $m^{-1/3}s$. The simulation time was 120 s for each case. The channel flow and runoff sheet flow coexisted: the runoff from the watershed surface was accumulated in the bottom of the watershed channel with a triangle-shaped cross section formed by the side slopes. Results of cases 2A and 2C are shown in **Figures 7** and **8**, respectively.

Figure 7 shows the comparison between the numerical solution and experimentally observed runoff hydrograph of Case 2A. The solution of the CCHE2D model agrees very well with the experimental results. The flow discharge increased continuously once the rain started. The peak discharge occurred at the time the rainfall stopped (at 45 s). Although the rising and the falling limbs of the hydrograph were

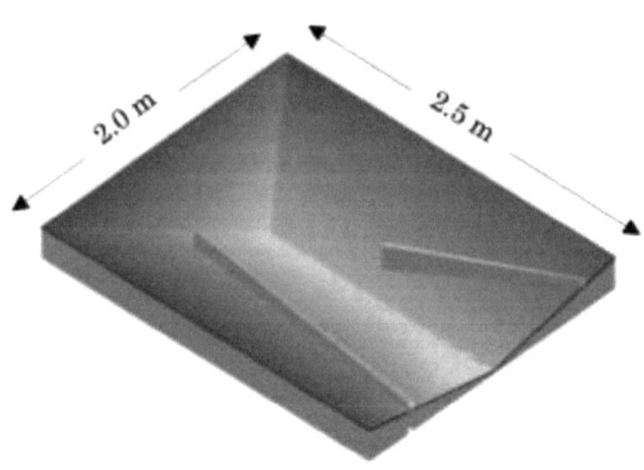

Figure 6.
Topography of the experimental watershed [30].

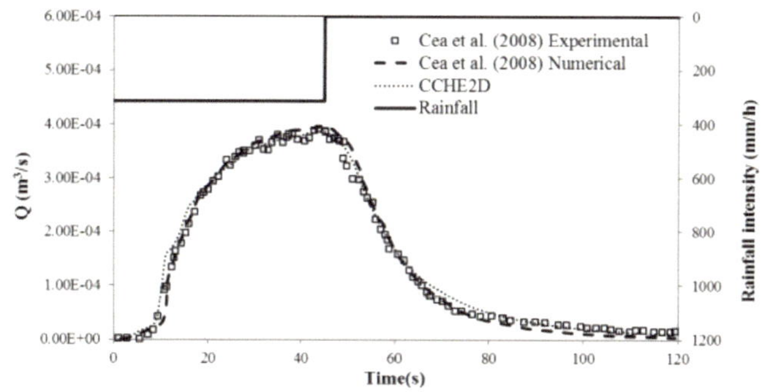

Figure 7.
Comparison of measured and simulated hydrographs using rainfall with one peak (Case 2A).

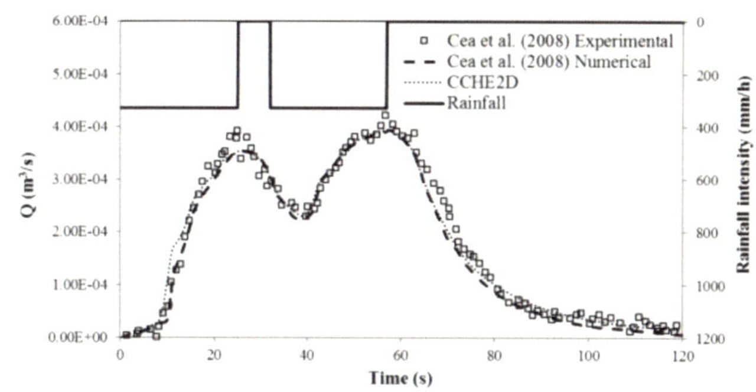

Figure 8.
Comparison of measured and simulated hydrographs using rainfall with two peaks (Case 2C).

slightly overestimated, the shape of the hydrograph and the peak discharge were aptly predicted.

Figure 8 shows the comparison between the numerical and experimental runoff hydrographs of Case 2C. The shape of the hydrograph was successfully predicted. The interval between the two rainfall peaks was 7 s. The first runoff peak discharge occurred at the time the rainfall stopped, at 25 s. The runoff discharge decreased for approximately 10 s and then increased. The second runoff peak discharge occurred at approximately 57 s. The simulated processes and the observed physical

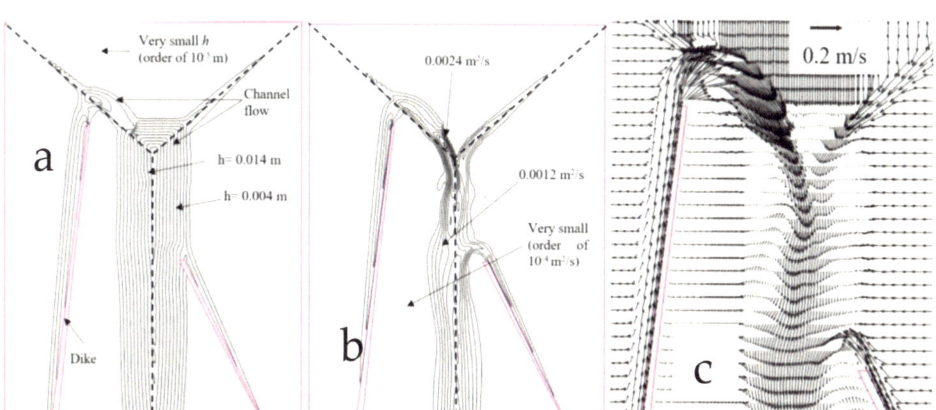

Figure 9.
Distributions of simulated (a) water depth contours (b) flow (unit discharge) distribution and (c) velocity vectors at t = *54 s for test Case 2C.*

processes showed a good general agreement; it also matched well with the model results of [30].

Figure 9 shows the simulation results at t = 54 s (the peak of the second rainfall) for Case 2C: (a) simulated water depth contour distribution, (b) simulated flow unit discharge pattern and (c) velocity vector distribution in the watershed. The distributions indicate how the overland sheet flow, under the influence of dikes and topography, concentrates into channels and flows out of the watershed. The flows over the slopes are sheet runoff, but complex recirculations are developed in the main channel. The water surface is no longer parallel to the bed surface. These flows cannot be represented by KW, DW, and SV models.

6. Application to a real watershed mathematical model

This section presents the application of CCHE2D to a sub-watershed of the Howden Lake watershed, an 18 km^2 agricultural watershed in the Mississippi River alluvial plain (**Figure 10**). In this region of low relief, watersheds are configured by farm fields drained by culverts, ditches, and intermittently flowing streams called bayous. During periods between runoff events, the channels contain standing water. The studied sub-watershed was upstream of a gaging station on an intermittently flowing bayou. The average annual precipitation in this region is about 1440 mm. Precipitation occurs as intense thunderstorms or low-intensity rains associated with major frontal movements. The latter type of events may stretch over several days of drizzle and sporadic showers. During growing seasons, channels experience some flow and stage fluctuation due to irrigation withdrawals and return flows.

Watershed topography was surveyed by airborne LiDAR with a 1.5 m horizontal resolution. The vertical accuracy was 15.0 cm RMSE or better. The watershed elevation ranges from approximately 43.89–48.99 m. A nearly uniform fine mesh (mesh spacing = 3.76–4.98 m) was generated for the simulation with the ditches and small streams between the plots further refined locally. Cultivated fields are

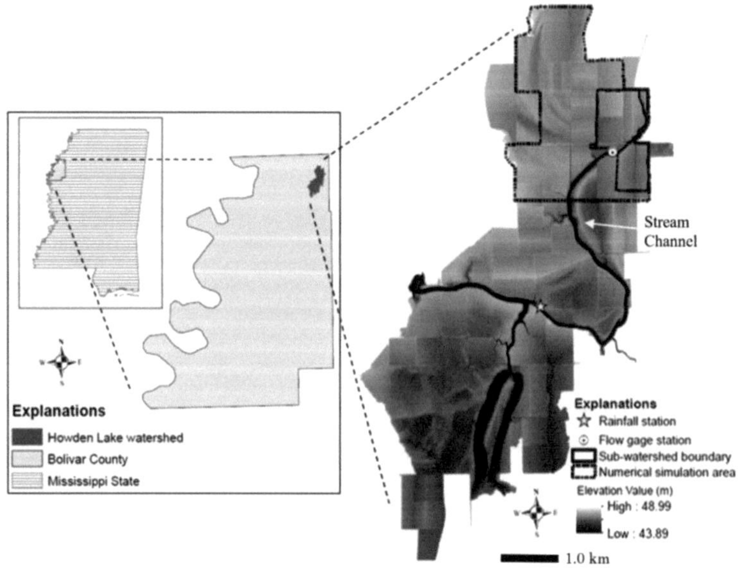

Figure 10.
Location and topography of the Howden Lake watershed. Dashed curve encloses the runoff simulation area, and the dark closed curve is the gaged watershed.

connected to the streams and ditches with drainage culverts, which often convey water from one sub-watershed to another. The locations of culverts in the study watershed were identified in a field survey and incorporated in the numerical mesh.

Soil data were obtained from the Soil Survey Geographic (SSURGO) database [47]. The watershed is covered mostly by soils with high clay content, which is typical of the region [48]. Infiltration is, therefore, negligible and was not considered in the simulation. Precipitation and flow stage data were measured by field instrumentation. Because the stream instrumented with the gage station has complex conditions, it was difficult to collect reliable velocity data during a rain event. Only stage data were available. As a result, the gage station does not have a discharge-stage rating curve. Development of a rating curve using simulations and measured data for this site would be helpful for understanding the hydrologic processes in these watersheds.

Because the Howden Lake watershed is of low-relief, it was often difficult to determine the boundaries between sub-watersheds in field surveys or on topographic contour maps. For example, the runoff from a piece of field may flow in two directions into two sub-watersheds, and the location of the divide line might be identified only from the runoff flow distribution during a simulation. Normally, the outline of a watershed is a given condition for a hydrology study. In this study, the exact boundary outline was not firmly established even after field surveys. A larger area containing the studied watershed was simulated, and the watershed boundary and area were finally defined by the simulated runoff and channel flow patterns. The boundary outline of the studied watershed (**Figure 11**) contributing to the gage was identified by visually checking the simulated overland flow directions of CCHE2D.

In the simulations, the streams and ditches between farming plots were represented using DEM elevations like flat surface areas. No channel networks were prescribed, but the simulated surface runoff flowed logically to the ditches and to the stream channel. No other watershed analysis tools were needed. Although the study results presented later are for this identified watershed, the spatial domain of numerical simulations was several times larger (**Figure 10**). The northern side of the stream channel had been blocked by farmers, so the overland flow from the watershed entered the stream in the middle and flowed in a southwesterly direction (**Figure 11**). The water from this identified watershed pasted the gage, while runoff from the region outside this watershed was discharged from the simulation domain via other ditches and streams. The area of this watershed, including cultivated land, drainage ditches and a stream segment, was found to be 973,700 m^2. In this area, the topographic elevation ranges from approximately 46.77–47.49 m in one plot and from 47.27 to 48.09 m in another. The mean slope of the fields is 0.0097 and 0.0098, respectively.

Several observed storm events were selected for the model application. To reduce minor losses of water due to evaporation, soil wetting and infiltration, etc., only large rain events were considered. The rainfall event in April 2011 (**Table 1**) was first used for simulation. **Figure 12a** shows the detailed ground elevation contour of a small simulation area (dashed rectangle area in **Figure 11**). The elevation of this area ranges from about 46.8 to about 47.4 m. **Figure 12b** shows the direction vectors of the runoff near the end of the simulation. Because the water is very shallow, the flow direction is highly affected by the ground topography. **Figure 12c** and **d** shows the direction vectors and water depth distribution at the peak time of the rainfall.

Although the variation of the bed surface topography is very small, the simulation shows how the runoff is controlled by microtopography (**Figure 12a**). At the peak time of the rainfall, the overall water depth in this area (**Figure 12d**) is much

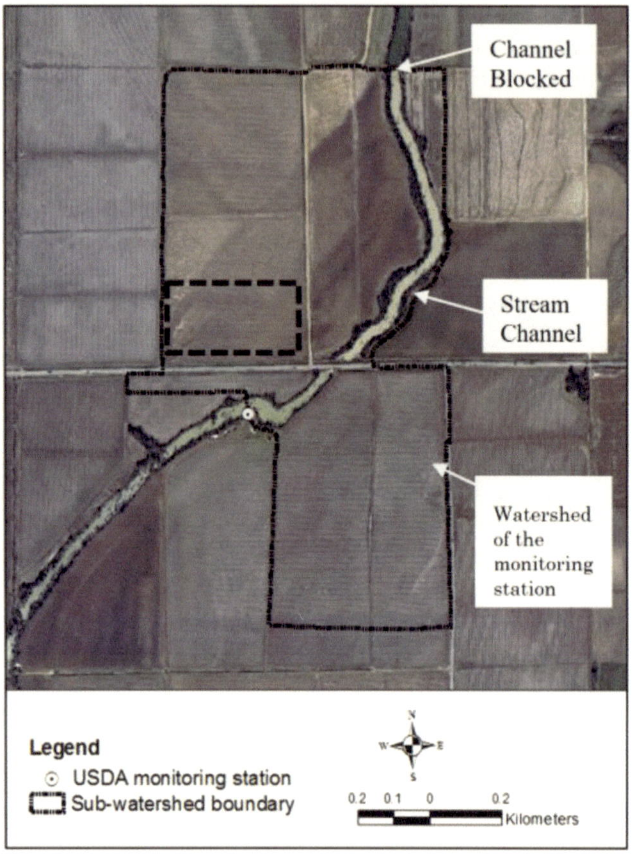

Figure 11.
*Numerical simulation identified watershed for the gage station. Simulation results in the dashed rectangle area are shown in **Figure 12**.*

Event	Measured rainfall (mm)	Runoff volume* (m^3)	z	r	L_0 (m)
4/27–4/28/2011	88.39	85,817	2.4	1.223	0.45
10/30–11/4/2013	53.59	52,182	1.9	4.24	0.78
11/21–25/2011	62.99	61,333	1.4	1.613	0.45
5/20–24/2013	48.77	47,483	1.0	1.436	0.59
9/25–27/2011	52.32	50,946	1.8	5.211	0.48

**Computed from the main bulk of the rain event.*

Table 1.
Parameters of selected runoff events for numerical simulations.

deeper, and the flow directions (**Figure 12c**) are less affected by the local microtopographic features. The flow on the right side of the domain is still sheet runoff under the KW condition; while on the left side, the water depth is more than 0.2 m, and the flow is no longer governed by the KW condition. This model provides the outflow hydrograph as well as the temporal and spatial distribution of the water depth and flow velocity, which can be used for studying soil erosion, agro-pollutant transport, and water quality.

The gage station (**Figure 11**) recorded the channel water surface elevation at regular time intervals, but velocities were generally too low for accurate

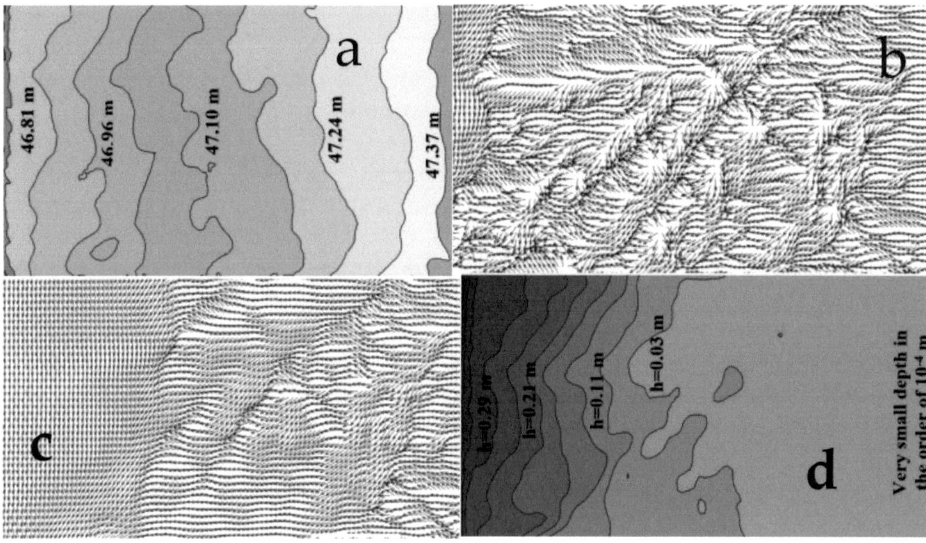

Figure 12.
Information and simulation results in an area indicated in **Figure 11** *in a dashed rectangle: (a) bed elevation contours, (b) velocity direction distribution near the end of the simulation, (c) velocity direction, and (d) the water depth distribution at the peak of the April 2011 rainfall.*

measurement, and therefore, water discharge was not measured. In order to better understand the watershed hydrology, a rating curve of the form:

$$Q = r(L - L_O)^z \tag{11}$$

was developed using simulated discharge, in which L is the measured water surface elevation, r and z are parameters, and L_O is the initial water surface elevation prior to a rainfall event. Eq. (11) has two unknown parameters, but there is only one relationship available for determining their values. The total volume of runoff, obtained by numerically integrating Eq. (11) in time, is equal to the rain volume, V_R:

$$\sum_i Q_i \Delta t = \sum_i r(L_i - L_O)^z \Delta t = V_R \tag{12}$$

in which L_i is the measured water surface elevation at the gage station. With Eq. (12) satisfied, values of r and z that best fit the shape of the discharge hydrograph computed using Eq. (11), and that of the numerical simulation, were determined for each event by trial and error.

Attempts were made to fit all simulated curves using a single set of values for r, z and a mean base stage L_O, but the result showed unacceptable discrepancies. L_O varied due to antecedent precipitation, downstream discharge control, sedimentation, and water usage between events. The range of L_O for the studied events is 0.33 m (**Table 1**). Given the complexities of the hydraulic regime in the water body, varying from standing to moving state and with varying downstream controls, variable rating curve parameters are sensible. Event-specific rating curve parameters are not ideal but are useful in a research context.

Manning's roughness coefficient (n) is a major factor in the determination of watershed runoff characteristics and generally reflects ground cover and management. The event on April 27–28, 2011 was used for initial calibration of Manning's coefficient. The studied watershed is cultivated with soybeans (*Glycine max* L. Merr.), corn (*Zea mays* L.), cotton (*Gossypium hirsutum* L.), and rice (*Oryza sativa* L.).

The sensitivity of the CCHE2D model in Howden Lake watershed to Manning's n was examined using a wide range of values from 0.030 to 0.30 $m^{-1/3}s$. Smaller Manning's n results in a higher runoff peak discharge and an earlier peak flow arrival time. A visual comparison of discharge hydrographs based on stage measurements and numerical simulation (**Figure 13**) indicates that n = 0.3 $m^{-1/3}s$ is the most appropriate choice for the overland runoff area because the peak times of these runoff events are consistent. Considering that the depths of the sheet runoff are much smaller than the microtopographic irregularities over the fields, the calibrated n represents not only the bed resistance but also form drags due to the microbed forms, crop residue, and vegetation. This n value agrees with the recent runoff studies [25, 31, 49] in cases of overland flows, including those in the Goodwin Creek Experimental Watershed in Northern Mississippi. There are numerous trees, bushes, and weeds growing along and within the channel, thus, n = 0.16 $m^{-1/3}s$ was used for the channel and kept unchanged for other rain event cases.

The total observed rainfall volume for the April 27–28, 2011 event (**Figure 13**) was approximately 86,000 m^3 (88.32 mm). The total simulated runoff volume is about 80,600 m^3 (83.78 mm), which is reasonable because the hydrograph recession limb extended past the simulation termination at 47 h. There were several small rain events that occurred before the event shown in **Figure 13**, so the runoff volume based on the observed water surface elevation may include recession of the earlier events.

Figure 14 compares the discharge hydrographs of several additional runoff events computed using Eq. (12) and that of the numerical simulations. The identified parameters for these events, r and z, are listed in **Table 1**. Events 9/2011 and 10/2013 have one major peak, while those of 11/2011 and 5/2013 each have two major peaks. The simulated hydrographs fit well with those computed using Eq. (11). The two rain peaks of the 5/2013 event were separated by about 2 h, but those of the 11/2011 event were separated by 15 h. The runoff of the 5/2013 event showed only one peak because the two rain peaks were very close, and the runoff peaks were superimposed. However, the temporal separation of the two peaks of the 11/2011 event was much longer. Therefore, the superimposed hydrologic response also displayed two peaks. These watershed responses were reproduced by the numerical simulations.

As noted above, the watershed has multiple field ditches that convey runoff into the channel (**Figures 10** and **11**). Ditch and channel flow were simulated together

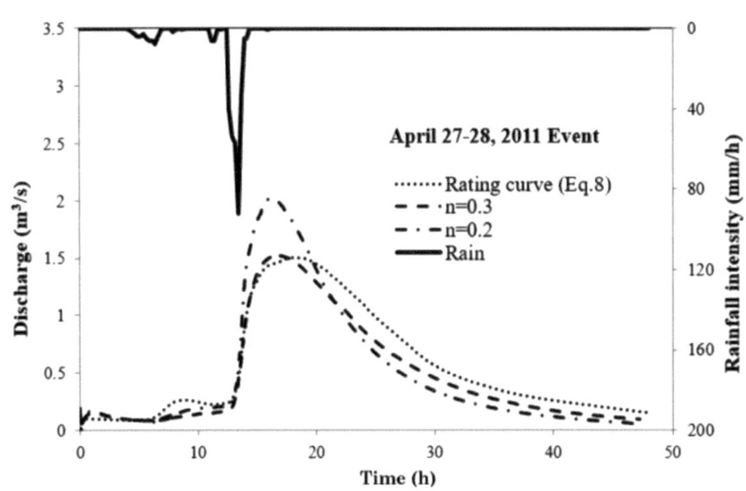

Figure 13.
Sensitivity of simulated hydrograph to Manning's coefficient.

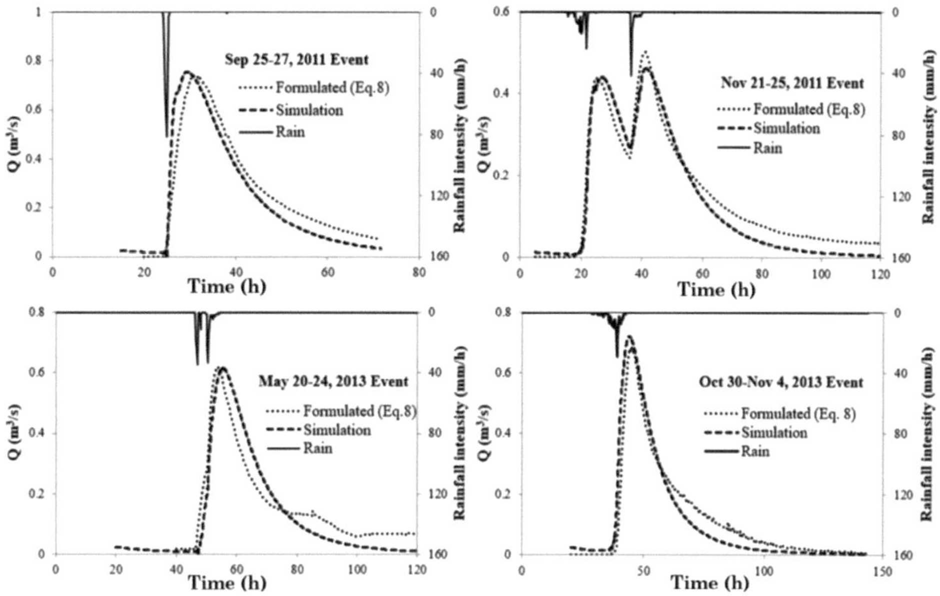

Figure 14.
Comparisons of simulated runoff and Eq. (11).

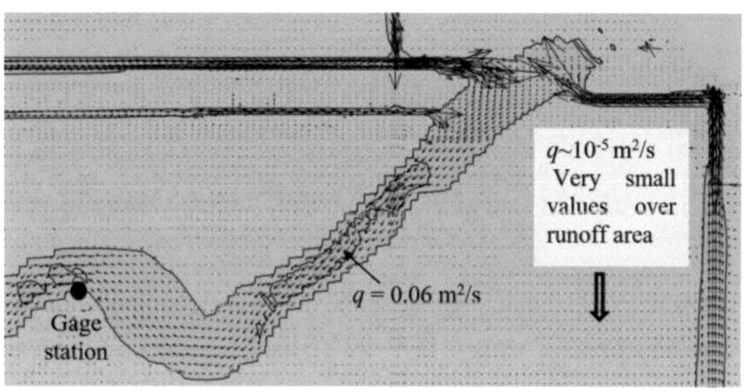

Figure 15.
Simulated flow in the network of drainage ditches and the stream in the watershed.

with the overland sheet runoff. **Figure 15** shows the simulated flows in the channel network of the watershed. The contours represent the distribution of the unit flow discharge. The vectors in the ditches and in the stream formed a channel network indicated by the large velocity vectors; those in the runoff area are too small to be seen. The flows in the stream are turbulent when the rainfalls are large. Because no velocity data were acquired, the simulated velocity results in the channel were not validated.

7. Conclusions

The numerical model CCHE2D was used to model sheet runoff from watersheds, large and complex enough to include both overland and channel flow processes. The model was systematically verified and validated using analytical solutions and experimental data due to steady and unsteady rainfall intensity, and applied to a real world watershed. Good agreement between the analytical solutions, experimental data, and numerical simulations were obtained. For the experimental cases involving complex watershed shapes, the numerical model has the ability to simulate runoff over the slope surfaces and the channel flows.

A numerical scheme was developed to correct the bilinear interpolation of the water surface elevation from its solutions at the staggered cell centers to the collocation nodes. The scheme was necessary and effective for obtaining good sheet runoff simulation results in watersheds with irregular topography. One would have to smooth the ground topography if a model requires the interpolation of water surface solution under this condition.

The model was applied to an agricultural watershed in the Mississippi River alluvial plain. It was useful to identify the boundary of the monitored watershed and develop the rating curve at the gage station of the watershed. Several significant runoff events were selected for simulation. Each of the simulated runoff hydrographs and the rating curves agreed well with those observed in the field. The sensitivity of the model to overland sheet flow friction was studied. An increase in the bed surface friction coefficient significantly diminishes the peak of runoff discharge, delaying its time of arrival. Values of n = 0.2–0.3 $m^{-1/3}s$ for overland flow were found to be adequate to best fit the numerical simulations and the observed data in the studied watershed. With a high-resolution mesh, the model can predict the complex surface runoff pattern over the agricultural land. Ditch and stream channels in the domain are a connected channel network. The model is able to simulate sheet runoff, turbulent channel flow, and their transitions seamlessly. The simulated hydrological processes for several storm events fit well to those observed at the gage station. The capability would be useful for studies related to soil erosion and agro-pollutant transport. The model is currently used for watershed applications without considering interception, evapotranspiration, and infiltration. Additional work is needed to further extend the research in these areas.

Acknowledgements

This work is supported in part by USDA Agriculture Research Service under the Research Project No. 6060-13000-025-00D (NCCHE) monitored by the USDA-ARS National Sedimentation Laboratory (NSL). Support is also in part by the Southeast Region Research Initiative (SERRI) project and the University of Mississippi (UM).

DOI: http://dx.doi.org/10.5772/intechopen.80214

Author details

Yafei Jia[1]*, Tahmina Shirmeen[1], Martin A. Locke[2], Richard E. Lizotte Jr.[2] and F. Douglas Shields Jr.[3]

1 National Center for Computational Hydroscience and Engineering, The University of Mississippi, MS, USA

2 Water Quality and Ecology Research Unit, USDA-ARS National Sedimentation Laboratory, MS, USA

3 Shields Engineering, LLC, MS, USA

*Address all correspondence to: jia@ncche.olemiss.edu

References

[1] Kirkby MJ, editor. Hillslope Hydrology. Chichester: John Wiley; 1978. p. 389

[2] Schmidt J. Soil Erosion: Application of Physically Based Models. Springer Verlag Berlin and Heidelberg GmbH Co KG; 2000. p. 318

[3] Liggett JA, Woolhiser DA. Difference solutions of the shallow water equations. Journal of the Engineering Mechanics Division, ASCE. 1967;**93**(2):39-71

[4] Cea L, Puertas J. Experimental validation of two-dimensional depth-averaged models for forecasting rainfall-runoff from precipitation data in urban areas. Journal of Hydrology. 2010;**382**: 88-102

[5] Kivva SL, Zheleznyak MJ. Two-dimensional modeling of rainfall-runoff and sediment transport in small catchments areas. International Journal of Fluid Mechanics Research. 2005; **32**(6):703-716

[6] Chow VT, Maidment DR, Mays LW. Applied Hydrology. New York: McGraw-Hill Book Company; 1988. p. 572

[7] Singh VP. Kinematic wave modelling in water resources: A historical perspective. Hydrological Processes. 2001;**15**(4):671-706

[8] Singh VP. Is hydrology kinematic? Hydrological Processes. 2002;**16**(3): 667-716. 1

[9] Morris EM, Woolhiser DA. Unsteady one-dimensional flow-over a plane: Partial equilibrium and recession hydrographs. Water Resources Research. 1980;**16**(2):355-360

[10] Book DE, Labadie JW, Morrow DM. Dynamic vs. Kinematic Routing in Modeling Urban Storm Drainage. In: Proceedings of the Second International Conference on Urban Storm Drainage. Yen BC editor. 14-19 June 1981. Urbana, IL, USA: University of Illinois at Urbana Champaign, 154-163

[11] Woolhiser DA, Liggett JA. Unsteady, one dimensional flow over a plane—The rising hydrograph. Water Resources Research. 1967;**3**(3):753-771

[12] Freeze RA. Mathematical models of hillslope hydrology. In: Kirkby MJ, editor. Hillslope Hydrology. New York: Wiley Interscience; 1978. pp. 177-225

[13] Cundy TW, Tento SW. Solution to the kinematic wave approach to overland flow routing with rainfall excess given by the Philip equation. Water Resources Research. 1985;**21**: 1132-1140

[14] de Lima JLMP, Singh VP. The influence of the pattern of moving rainstorms on overland flow. Advances in Water Resources. 2002;**25**:817-828

[15] Rousseau M, Cerdan O, Delestre O, Dupros F, James F, Cordier S. Overland flow modelling with the shallow water equation using a well-balanced numerical scheme: Adding efficiency or just more complexity? 2012. Available from: https://hal.archives-ouvertes.fr/hall-00664535

[16] Du JK, Shunping X, Youpeng X, Xu C, Singh VP. Development and testing of a simple physically-based distributed rainfall-runoff model for storm runoff simulation in humid forested basins. Journal of Hydrology. 2007;**336**:334-346

[17] Zhang W, Cundy TW. Modeling of two-dimensional overland flow. Water Resources Research. 1989;**25**(9): 2019-2035

[18] Gottardi G, Vinutelli M. An accurate time integration method for simplified

overland flow models. Advances in Water Resources. 2008;**31**:173-180

[19] Esteves M, Faucher X, Galle S, Vauclin M. Overland flow and infiltration modeling for small plots unsteady rain: Numerical results versus observed values. Journal of Hydrology. 2000;**228**:265-282

[20] Howes DA, Abrahams AD, Pitman EB. One- and two-dimensional modelling of overland flow in semiarid shrubland, Jornada basin, New Mexico. Hydrological Processes. 2006;**20**:1027-1046

[21] Liu QQ, Chen L, Li JC, Singh VP. Two-dimensional kinematic wave model of overland-flow. Journal of Hydrology. 2004;**291**(1–2):28-41

[22] Costabile P, Costanzo C, Macchione F, Mercogliano P. Two-dimensional model for overland flow simulations: A case study. European Water. 2012;**38**: 13-23

[23] Nunoz-Carpena R, Miller CT, Parsons JE. A quadratic Petrov-Galerkin solution for kinematic wave overland flow. Water Resources Research. 1993; **29**(8):2615-2627

[24] Venkata RK, Eldho TI, Rao EP. A diffusion wave based integrated FEM-GIS model for runoff simulation of small watersheds. Journal of Water Resource and Protection. 2009;**1**:391-399. Published online December 2009. DOI: 10.4236/jwarp.2009.16047

[25] Singh J, Altinakar MS, Ding Y. Numerical modeling of rainfall-generated overland flow using nonlinear shallow-water equations. Journal of Hydrologic Engineering. 2104:1943-5584. https://doi.org/10.1061/(ASCE) HE.1943-5584.0001124

[26] Shirmeen T, Jia Y, Locke MA, Lizotte Jr. RE. Numerical modeling of rain induced overland flows. In: World Environmental and Water Resources Congress, 2015; 17-21 May 2015; Austin, Texas. 2015

[27] Iwagaki Y. Fundamental Studies on the Runoff Analysis by Characteristics. Vol. 10. Japan: Bulletin - Kyoto University Disaster Prevention Res. Inst.; 1955. pp. 1-25. 2

[28] Govindaraju RS, Kavvas ML, Jones SE. Approximate analytical solutions for overland flows. Water Resources Research. 1990;**26**(12):2903-2912

[29] Singh VP. Kinematic Wave Modeling in Water Resources: Surface Water Hydrology. Chichester: John Wiley and Sons Ltd.; 1996

[30] Cea L, Puertas J, Pena L, Garrido M. Hydrologic forecasting of fast flood events in small catchments with a 2D-SWE model. Numerical model and experimental validation. In: World Water Congress 2008; 1-4 September 2008; Montpellier, France

[31] Downer CW. Demonstration of GSSHA Hydrology and Sediment Transport at the Goodwin Creek Experimental Watershed. ERDC TN-SWWRP-08-x. Vicksburg, MS: U.S. Army Engineer Research and Development Center; 2008 Available from: https://swwrp.usace.army.mil/

[32] Yeh GT, Shih DS, Cheng JRC. An integrated media, integrated processes watershed model. Computers & Fluids. 2011;**45**(2011):2-13

[33] Graham DN, Butts MB. Flexible, integrated watershed modelling with MIKE SHE. In: Singh VP, Frevert DK, editors. Watershed Models. CRC Press; 2005. pp. 245-272. ISBN: 0849336090

[34] Ewen J, Parkin G, O'Connell PE. SHETRAN: Distributed river basin flow and transport modeling system. Journal of Hydrologic Engineering, ASCE. July 2000;**5**(3):250-258

[35] Vieira DA, Wu W. One-Dimensional Channel Network Model CCHE1D Version 3.0: User's Manual (Report NCCHE-TR-2002-2). Mississippi, USA: National Center for Computational Hydroscience and Engineering, The University of Mississippi; 2000

[36] Jia Y, Wang SSY, Xu YC. Validation and application of a 2D model to channel with complex geometry. International Journal of Computational Engineering Science. 2002;**3**(1):57-71

[37] Jia Y, Zhang Y, Wang SSY. Simulating bank erosion process using a depth averaged computational model. In: 7th Symposium on River, Coastal and Estuarine Morphodynamics (RCEM 2011), 2011; 5-8 September 2011; Beijing, China, CD-ROM

[38] Jia Y, Chao XB, Zhang YX, Zhu TT. Technical Manual of CCHE2D, V4.1, NCCHE-TR-02-2013. The University of Mississippi; 2013

[39] Jia Y, Altinakar M, Chao X, Zhang Y. Numerical simulations of spilled coal ash in the Dan River and the environmental impact of the incident. In: World Environmental and Water Resources Congress. 2016. pp. 114-125. DOI: 10.1061/9780784479865.012

[40] Ding Y, Wang SSY, Jia Y. Development and validation of a quasi-three dimensional coastal area morphological model. ASCE, Journal of Waterway, Port, Coastal, and Ocean Engineering. ASCE. 2006;**132**(6):462-476

[41] Wang SSY, Roche PJ, Schmalz RA, Jia Y, Smith PE. Verification and Validation of 3D Free-Surface Flow Models. American Society of Civil Engineering; 2008. ISBN: 968-0-7844-0957-2

[42] Jia Y, Wang SSY. Numerical model for channel flow and morphological change studies. Journal of Hydraulic Engineering, ASCE. 1999;**125**(9): 924-933

[43] Jia Y, Shirmeen T. Simulation of rainfall-runoff process in watersheds using CCHE2D. In: 12th International Conference on Hydroscience & Engineering: Hydro-Science & Engineering for Environmental Resilience; 6-10 November 2016; Tainan, Taiwan

[44] Singh VP, Regl RR. Analytical solutions of kinematic equations for erosion on a plane. I. Rainfall of indefinite duration. Advances in Water Resources. 1981;**6**(1):2-10

[45] Singh VP. Analytical solutions of kinematic equations for erosion on a plane. II. Rainfall of finite duration. Advances in Water Resources. 1983; **6**(2):88-95

[46] Morgali JR, Linsley RK. Computer analysis of overland flow. Journal of Hydrological Division, Proc. American Society of Civil Engineers. 1965;**HY3**: 81-100

[47] Soil Survey Staff, Natural Resources Conservation Service, United States Department of Agriculture, Web Soil Survey. Available from: https://websoilsurvey.nrcs.usda.gov/ [Accessed: 10-04-2017]

[48] Fisk HN. Geological Investigations of the Alluvial Valley of the Lower Mississippi River. Vicksburg, Mississippi: U.S. Army Corps of Engineers, Mississippi River Commission; 1944. p. 78

[49] Kalyanapu AJ, Burian SJ, McPherson TN. Effect of land use-based surface roughness on hydrologic model output. Journal of Spatial Hydrology. 2009;**9**(2):51-71. Fall 2009

Section 3

Assessment of Soil Erosion Risk

Chapter 4

Evaluating Differences of Erosion Patterns in Natural and Anthropogenic Basins through Scenario Testing: A Case Study of the Claise, France and Nahr Ibrahim, Lebanon

Mario J. Al Sayah, Rachid Nedjai, Chadi Abdallah, Michel Khouri, Talal Darwish and François Pinet

Abstract

This study assessed soil erosion risks of two basins representing different geographical, topographical, climatological and land occupation/management settings. A comparison and an evaluation of site-specific factors influencing erosion in the French Claise and the Lebanese Nahr Ibrahim basins were performed. The Claise corresponds to a natural park with a flat area and an oceanic climate, and is characterized by the presence of 2179 waterbodies (mostly ponds) considered as hydro-sedimentary alternating structures, while Nahr Ibrahim represents an orographic Mediterranean basin characterized by a random unequal land occupation distribution. The Claise was found to be under 12.48% no erosion (attributed to the dense pond network), 65.66% low, 21.68% moderate and 0.18% high erosion risks; while Nahr Ibrahim was found to be under 4, 39.5 and 56.4%, low, moderate and high erosion risks, along with 66% land degradation determined from the intersection of land capability and land occupation maps. Under the alternative scenario for the Claise where ponds were considered dried, erosion risks became 1.12, 0.52, 76.8 and 21.56%, no erosion, low, moderate and high risks, respectively. For Nahr Ibrahim, and following the Land Degradation Neutrality intervention, high erosion risks decreased by 13.9%, while low and moderate risks increased by 3 and 10.8%.

Keywords: erosion, LDN, land degradation, ponds, Mediterranean climate, oceanic climate

1. Introduction

Soil erosion is considered as the most amplified manifestation of land loss worldwide. It has become one of the most pressuring global problems facing sustainable development at rates exceeding pedogenesis by 10–40 times [1].

According to Lal [2], a worldwide area of 1094 million ha is subject to soil erosion, of which 751 million ha have been severely eroded. As a result of soil erosion, significant declines in land quality due to the loss of the much needed fertile topsoil layers used for agriculture and for providing primary eco-services have been reported [3] particularly in arable lands whose decline accounts for losses in the order of 400 billion US dollars/year globally [4]. From the various erosion forms, water erosion is considered as the most problematic, due to the increase in its extent and intensity, leading to deleterious losses in land capital and environmental sustainability [4, 5]. Europe and the Mediterranean particularly are significantly affected by this process [6, 7] where in Europe, soil erosion is one of the most threatening challenges for soil resources causing losses of 3–40 t/ha/year [8] while in the Mediterranean region, particularly in its Middle Eastern and North African parts [9, 10], soil erosion rates have significantly surpassed Mediterranean pedogenesis rates [11, 12].

In Europe, soil loss can be attributed primarily to water erosion due to climate (abundant rainfall), soil management practices and agrarian intensification coupled to unsustainable practices such as overgrazing [13]. In the Mediterranean region on the other hand, factors are much more complex due to the pronounced rainfall variability and heterogeneity of site-specific characteristics [14] even within the same landscape. As a result of weakly resistant pedology [15], unequal and random land use/land cover distribution [16] occurring due to the absence of governance, management plans and restraints [17], low precipitations, erratic intense rain episodes, prolonged droughts, steep slopes and increasing anthropogenic effects [18], soil erosion has reached an irreversible state in some regions, while in others erosion has ceased because no more soil is left to erode [13]. Consequently, soil erosion has led to the process of land degradation causing significant loss of land capital [19], thereby threatening food security and sustainable development in the region [20]. For that purpose, a simultaneous assessment englobing both soil erosion and land degradation must be carried out. Nevertheless, this task is contested by several factors namely the non-uniqueness of definitions of the process [21], the existence of unmeasurable interdependent driving factors [22] and the absence of clear methodological or application workflows [23].

This deteriorating state of soil erosion in Europe has led to the development of the European common framework for the Thematic Strategy on Soil Protection and the Common Agricultural Policy that highlight the need to protect European soils to reduce soil erosion [4, 24]. In contrast, the Mediterranean basin still lacks concrete and direct policies or legislations targeting soil erosion [25] due to contested definitions of land loss in the region [26]. Under any circumstance and prior to treating soil erosion, assessing its extent and identifying hotspots are required [13, 27]. However, this assessment is not an easy task given the heterogeneity and large spatial/temporal variability of its driving factors [28, 29], particularly in Mediterranean landscapes [30] that are characterized by a complexity of slope, climate and land occupation factors [10].

The process of soil erosion is attributed to various interdependent driving factors, notably climate, pedologic properties, topography and vegetation cover [31]. Despite being a natural process at its origin, soil erosion has significantly increased as a result of anthropogenic activity [32], where land use and land cover changes have become the main drivers of soil erosion [29] combined to soil management and conservation strategies [33]. When considering soil erosion, a multi-scale problem is at hand due to the role and status of soil erosion in several environmental, socioeconomic and developmental processes, often causing a cascade of direct on-site and indirect off-site effects. Under the environmental scope, soil erosion is considered as the main form of soil loss leading to negative impacts on water quality, biodiversity, organic carbon stocks and eco-services [24]. At the socio-economic scale, soil erosion has become one of the governing factors in land use allocation, notably under the scope of agriculture as function of market economy [34], where increasing needs for increased productivity led to significant removal of natural cover for agricultural expansion rendering large areas vulnerable to soil erosion [33]. At the developmental scale, soil erosion has caused notable declines in the productive capacity of lands, often leading them to become unproductive, ultimately resulting in agrarian abandonment [1]. The latter, in turn, causes an amplification of erosion due to increased exposure of soil to water [35], thus promoting loss of land and soil resources both quantitatively and qualitatively. Collectively, the previously cited factors culminate not only to create short-term losses in agricultural productivity [1], but also to affect long-term food security [36], thus imposing challenges for achieving sustainable development [37, 38].

Further, soil erosion forms the head component of van Rijn's [39] sedimentary cycle, consisting of erosion, transport and deposition, rendering it partly responsible for shaping the hydromorphological aspects of landscapes along with surface runoff, sediment transport, baseflow and stream discharge [40]. Given the status of soil erosion as the head of the sediment transport chain, changes of soil erosion are capable of causing a cascading effect influencing the whole cycle and ultimately modifying both the hydro-sedimentary response and equilibrium of basins, thus creating challenges for watershed managers [41].

Studies regarding soil erosion have received growing interest under different approaches; these have led to the development of several models for estimating erosion [42] of which the USLE [43], MUSLE and RUSLE [44] are some of the most basic yet widely used models. Other models such as EUROSEM [45], WEPP [46], CORINE [47], TOPOG [48] and SedNet [49] have also been employed at different scales and study areas with various degrees of success. Ref. [50] summarized a number of applied approaches for studying erosion that can be grouped under: (a) use of models (e.g., [12, 42, 51]), (b) erosion plot data for direct in-situ measurements (e.g., [52]) and (c) by means of measuring sediment yield (e.g., [53]) since the latter is the net product of soil erosion [54]. Among these various methods, the use of models has been deemed to be the best given its efficiency, not only for displaying current conditions but also for revealing changes resulting from alternative simulations presenting changes of natural conditions [55] in addition to overcoming the problems of field measurements and logistics.

For erosion assessment, the basin scale is considered as most suitable given its capacity to reveal anthropogenic-interference effect [56] and due to the fact that soil erosion is one of the most pronounced problems in basins posing a considerable challenge for hydrologists and basin managers [41]. Given the scope of this study for comparing natural and managed basins having different natural contexts under different land occupation and managed settings, the French Claise and Lebanese Nahr Ibrahim basins are chosen as study areas for establishing a comparative framework between two different geographical and management contexts.

The Claise basin is one of the several basins corresponding to the French Brenne Regional Natural Park. The latter is an international heritage area housing a large number of ponds in its premises, nearly 4500, of which 2179 are in the Claise [57]. It is chosen as a representative of Northern European basins which are often covered by a prevalent number of ponds. In France particularly, three main pond density zones are present; these are the Sologne region, Brenne (Centre France) and Dombes (Eastern France). Ponds are considered to be one of the most important hydro-sedimentary modifying manmade structures [58] that possess an aggregative effect far more important than larger water bodies [59] on altering the regime of basins they take part of. Therefore, in response to the recommendations of the Directive-Cadre

Européenne sur l'eau (DCE) [60], regarding the importance of understanding the impact of hydromorphological factors on watershed processes, the Claise basin which takes part of the Brenne Natural Regional Park is chosen as the natural watershed of this study. In contrast, the Nahr Ibrahim basin represents the managed basin of this study. It is a Lebanese basin known for excessive erosion rates [12] that have led to significant land degradation [61] and landslides [62] coupled to a typical Mediterranean unequal land occupation distribution that has expanded due to the absence of land use planning [20].

The workflow of this chapter consists of using the CORINE erosion model [48] given its relative accuracy with respect to simple data requirements consisting of climate, slope, soil properties and vegetation cover, and its widespread application [63]. Erosion assessment in the Claise basin serves to respond to DCE recommendations for assessment of the effect of hydromorphological altering structures on basins. For Nahr Ibrahim, the CORINE model serves as a tool for mapping land degradation as function of soil erosion. Following the establishment of both actual soil erosion maps, a comparison between the natural and managed settings allows the assessment of the impact of land occupation and management on erosion risks.

Given the flexibility of the CORINE model incorporating both natural (slope, pedology and climate) and vegetation cover (human controlled), alternative vegetation covers for both basins were used to re-assess changes in erosion patterns and risks. This step was performed to pinpoint the impact of ponds on erosion patterns of the Claise basin and to prospect the efficiency of the Land Degradation Neutrality (LDN) concept for erosion reduction through land use planning [64]. LDN is defined by the United Nations Convention to Combat Desertification (UNCCD), [65], to be "a state whereby the amount and quality of land resources necessary to support ecosystem functions and services and enhance food security remain stable or increase within specified temporal and spatial scales and ecosystems." LDN aims to halt ongoing losses by land degradation. Unlike past approaches, LDN creates a target for land degradation management by means of a dual phased approach containing measures to avoid or reduce land degradation as a first phase. The second phase presents a combination with the first where specific applications to reverse or to treat past degradation are employed in order to rehabilitate degraded zones. Therefore, the concept of neutrality involves counterbalancing losses and equivalent gains. However, many factors enter in the estimation of losses including the effects of planning decisions (e.g., granting permits for open-cut mining), the effects of past and previous decisions (e.g., continuation of agricultural practices known to deplete soil carbon) and mostly the natural drivers of land degradation (e.g., impacts of drought, wildfire) [67].

Ideally, the most effective strategy would be to take immediate action to prevent land degradation where non-degraded lands are at risk. For effective implementation, it is important to consider the resilience of the counterbalancing intervention over the long term, the potential impacts of climate change and the likely trade-offs between ecosystem services. For these reasons, the proposed land use scenario for the Nahr Ibrahim basin consists of a realistic plan accounting for the trade-off between natural resources and the need to promote sustainable urban development. This task is achieved following the LDN's "soil" indicators of land use/land cover change and soil organic C stocks in analogy to the work done by Al Sayah et al. [66] and in response to the LDN hierarchy involving three actions in descending order of importance: avoid, reduce and reverse.

Through this study, the comparative land occupation framework in addition to the alternative modeling approach aims to provide an understanding regarding the relationship between land occupation (as land use/land cover and management) and soil erosion, as well as integrating soil erosion as part of land planning.

2. Geographical context and site-specific description of the test-site basins

2.1 The Claise basin: a particular mosaic under a natural setting

The Indre section of the Claise basin (46° 56′ 23.89″ N and 1° 31′ 32.61″ E) is one of the three basins corresponding to the Brenne Regional Natural Park. The 1760 km^2 park is located in the French Centre-Val-de-Loire region and is renowned as the land of thousand ponds due to the presence of 4500 ponds extending in a natural landscape mosaic [57]. A large number of these water bodies are located in the corresponding section of Claise basin (2179 ponds) that describes an area of 707 km^2 [67] (**Figure 1**). These are speculated to be one of the key feeding sources of the 87.6-km-long Claise River (Rougé (1927) in [68]) described by an average flow of 4.50 m^3/s and originating at 146 m of altitude [69] with three main tributaries: the channel of the Five Bonds (or Blizon), the Yoson and Suin Rivers. Despite the proficient presence of water bodies within, the Claise basin is described by a poorly organized and extensively fragmented hydrological network [70]. Since the study area takes part of a national park, the land occupation pattern of the Claise basin has remained relatively unchanged for the last 19 years except for pond proliferation. The land occupation setting of the Claise consists mainly of a homogeneous interlocking mosaic of abundant grasslands, agricultural areas and forests as opposed to a very low urban occupation [69]. The climate of the Claise basin mainly

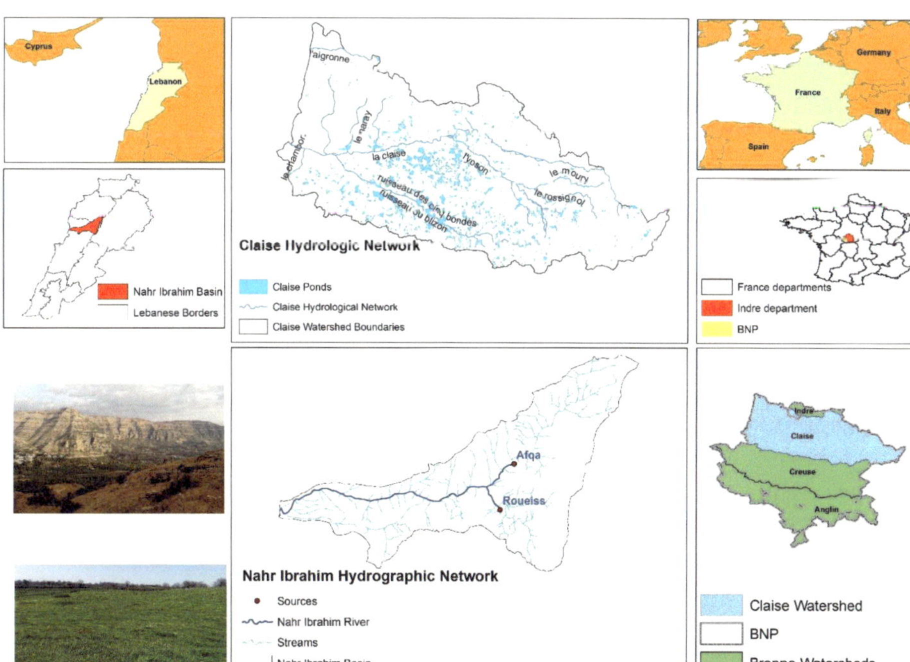

Figure 1.
Study area description, BNP: Brenne Regional Natural Park.

corresponds to the degraded oceanic continental climate with high oceanic influence having annual average temperatures of 11°C, 8–14 days of temperatures below −5°C and annual cumulative precipitations in the order of 700 mm [71]. However, Nedjai et al. [57] have shown that the pond dense zone possesses the ability to create a local microclimate quite different from its surrounding. In terms of topography, the Claise basin can be described as a flat area with an altitude range of 76–181 m. According to Fischer et al. [72], six soil groups are present in the basin; these are in descending order of spatial coverage: Luvisols, Podzols, Leptosols, Cambisols, Fluvisols and Arenosols. According to Barrier and Gagnaison [73], the geological setting is dominated by Cenomanian, Jurassic and clay deposits and was completed at the end of the Tertiary era.

As a result of its poor hydrographic network, quasi-impermeable pedological setting, litho-stratigraphic composition, flat topography and abundant rainfall, stagnation of incoming water in the basin resulted in the formation of ponds [57, 74]. However, the proliferation of ponds in great numbers is not only due to natural origins, but also a translation of significant anthropogenic interference to overcome economic restraints imposed by the challenging soil productive capacity for use for extensive aquaculture [57, 75]. Despite the proficiency of aquaculture in the region, the Brenne Regional Natural Park displays a population density of 17.9 inhabitants/km^2, which has been considered as one of the lowest in the Région Centre [76] and has been engaging in decreasing trends since the year 2006 [77]. This state leads to a population exodus in the study area, thus constricting further the presence and associated impact of anthropogenic activity.

Overall, the presence of a dominantly natural vegetated land cover and the absence of sloping areas generally imply a low erosive setting. However, given the questions raised regarding the impact of ponds, known to be modifiers of the hydro-sedimentary response of basins, particularly due to their presence in significant numbers and their position as a chain setting, this basin was chosen for investigation of the pond-impact on basin erosion risks.

2.2 The Nahr Ibrahim basin: a representative Mediterranean basin

The Nahr Ibrahim basin is one of the 11 coastal basins of Lebanon. It describes an area of 309 km^2 accounting for 3% of the country's area between 36° 2′ 46″ E, 34° 12′ 46″ N and 35° 38′ 35″ E, 33° 59′ 36″ N [62] and represents one of the most important Lebanese basins. The basin houses the perennial Nahr Ibrahim River, one of the 17 rivers protected by the Lebanese Ministry of Environment [78] given its biological and ecological significance and its role as a vital input for the local economy [79], primarily for agricultural irrigation, freshwater supplies and eco-tourism services [80]. The basin is characterized by a rich hydrological network consisting of several effluents feeding the 27-km-long river that originates from the Afqa and Roueiss springs at an altitude of 1200 m and 1265 m, respectively [62], and flows at 507 million m^3/year [81]. A typical heterogeneous Mediterranean basin land occupation pattern consisting of a heavily urbanized lower part, a semi-natural middle section and a mountainous upper basin accounting for nearly 60% of the basin is observed within. As many other regions of Lebanon, land occupation dynamics have occurred under a lack of governance, regulations, restraints and management plans [17] leading to an unequal repartition in the same landscape, thus giving rise to a heterogeneity of basin processes within. A typical Mediterranean climate showing increasing tendency toward prolonged droughts and more erratic intense rainfall events dominates the study area. Precipitations occur in the form of rainfall ranging from 900 mm to over 1400 mm, while in the upper mountainous part, snowfall is prevalent during the November–March period with a snow cover

Parameters	Claise basin	Nahr Ibrahim basin
Climate	Degraded oceanic	Mediterranean
Hydrological network	Severely fragmented, characterized by the presence of ponds in great numbers	Rich
Topography	Flat	Heterogeneous, characterized by steep slopes
Geology	Dominated by Cenomanian, Jurassic and clay deposits	Dominated by Cenomanian, Jurassic and Quaternary deposits
Pedology	Quasi-impermeable soil groups	Permeable soils with heterogeneous distribution
Land use/Land cover	Dominantly natural with the presence of manmade ponds in large numbers **Homogeneous**	Mountainous upper portion; middle region with a diversity of superficial lands exploited with urban zones, agricultural fields planted with fruit trees, pasture with low vegetation and forestry; heavily urbanized lower region **Heterogeneous**
Anthropogenic pressure	Light showing minor increases	Increasing

Table 1.
Comparison of study area characteristics.

often lasting until late summer [62]. Geomorphologically, the basin corresponds to a mountainous area characterized by a varied topography consisting of hills and valleys with an upward slope gradient of nearly 20–25 m/km, along with a moderately sharp surface relief extending between the coast and 2600 m of altitude [61]. According to Darwish et al. [82], the Nahr Ibrahim basin is comprised of 11 soil groups in descending order of spatial coverage. Soil Associations[1], Leptosols, Andosols, Regosols, Anthrosols, Arenosols, Luvisols, cliffs, Cambisols, Gleysols and Fluvisols. According to Dubertret [83], the geology of the basin is presented by eight rock units dominated by Cenomanian carbonate rocks (70%) followed by the Jurassic (20%), with outcropping stratigraphic sequences revealing rock formations spanning from the Middle Jurassic to the recent epoch. Socioeconomically, and as other regions of Lebanon, the Nahr Ibrahim basin presents a densely populated lower portion corresponding to its coastal area in contrast to a less populated upper mountainous region [10]. In addition to urbanization in its lower part, the Nahr Ibrahim basin suffers from intensive industrial development [84] as opposed to a much less populated mountainous upper part.

As a result of its complex topography, abrupt climatic conditions and pedological composition, the Nahr Ibrahim basin has been reported by Abdallah and Faour [62] to be a region of intensive landslides that cover up to 7.6 km^2 of its area due to the dominance of Leptosols extending over Cenomanian (C4) and Jurassic (J4) formations, generally found over karstic and sloped areas, thus rendering them vulnerable to erosion. Further, as a result of extensive anthropogenic activity, the basin has been reported to be an area of intensive sloping runoff with increasing vulnerability to erosion [12] in addition to increasing trends of land degradation [61], thus making it a suitable target for this study.

Since a comparative framework is targeted in this study, **Figure 1** presents the settings of both study areas, while **Table 1** presents a general comparison.

[1] Calcaric Leptosols, Haplic Leptosols, Skeletic Regosols, Leptic Luvisols and Lithic Luvisols.

3. Methodological workflow and theoretical aspects of the study

3.1 The CORINE erosion risk model framework: basis and concepts

Despite the prevalence of soil erosion assessment models, data availability limits the choice of sought models; therefore, given the data-sparse nature of the Nahr Ibrahim basin and the absence of quantitative soil loss studies in the country [42], the use of a process-based model is not possible. Therefore, a robust reliable model with relatively simple data requirement is sought. Accordingly, the semi-qualitative empirical CORINE erosion model has been chosen given its capability of accurately predicting the spatial distribution of erosion risks with relatively simple data requirement and ease of parameterization [63]. Despite its empirical nature which may provide it with an accuracy less than that of physical or process-based models, the CORINE model was chosen since empirical models are of adequate use for soil conservation studies [85]. Further, several successful applications of the model have been documented in different regions of the world (e.g., [86–88]), therefore giving it adequate reliability, particularly for the Mediterranean and data-sparse regions [89].

For erosion risk assessment using the CORINE model, several factors are required. These, according to Vertessy et al. [48], are:

1. *soil erodibility*, computed from three attributes—soil texture where fine particle fractions are more readily removed than coarser fractions [90], soil depth where deeper soils resist erosion as function of higher water-holding capacities [48] and stoniness given their protective role in the pre-surface runoff stage [87];

2. *soil erosivity*, computed from two climatic indices—the Modified Fournier Index (MFI) to determine rainfall variability [91] and the Bagnouls-Gaussen aridity Index (BGI) [92] to reveal the possibility of abrupt short-storm events during normally dry seasons [48] leading to intensive erosion;

3. *topography*, obtained through slope angle calculation given its pronounced effect on soil erosion, particularly when a certain critical threshold is exceeded [48];

4. *vegetation cover*, obtained from Land Use and Land Cover (LU/LC) maps given their effect on soil fixation via their roots and by reducing rainfall splash effect [5, 93].

The respective input layers were extracted from the databases and inputted into the Raster Calculator tool of ArcGIS for computation and application of the basis, equations and workflow for the CORINE presented in **Figure 2**. Each index was computed after classification into the corresponding CORINE categories, into the erodibility, erosivity and topography components which in turn are part of the potential soil erosion risk formula (**Figure 2**). Having obtained the potential soil erosion risk map, overlaying the vegetation cover layer allowed the computation of the actual soil erosion risk maps. Through scenario testing using a study adapted vegetation cover as an alternative input to the CORINE model, land degradation under the form of soil loss (here erosion) was determined. By quantitatively determining erosion risks using equations presented in **Figure 2**, an accurate representation of soil loss by erosion under current conditions is obtained. This step in turn serves as a reference or a baseline indicator for comparison with alternative scenarios. In the case of Nahr Ibrahim, for elaboration of measures to counterbalance the negative effects of land degradation, and balance land losses by land gains through application of the LDN concept, the CORINE model was used to reveal changes in erosion risks after LDN implementation. The latter is a new concept proposed in 2015 by the UNCCD to protect stable lands, halt ongoing degradation and restore

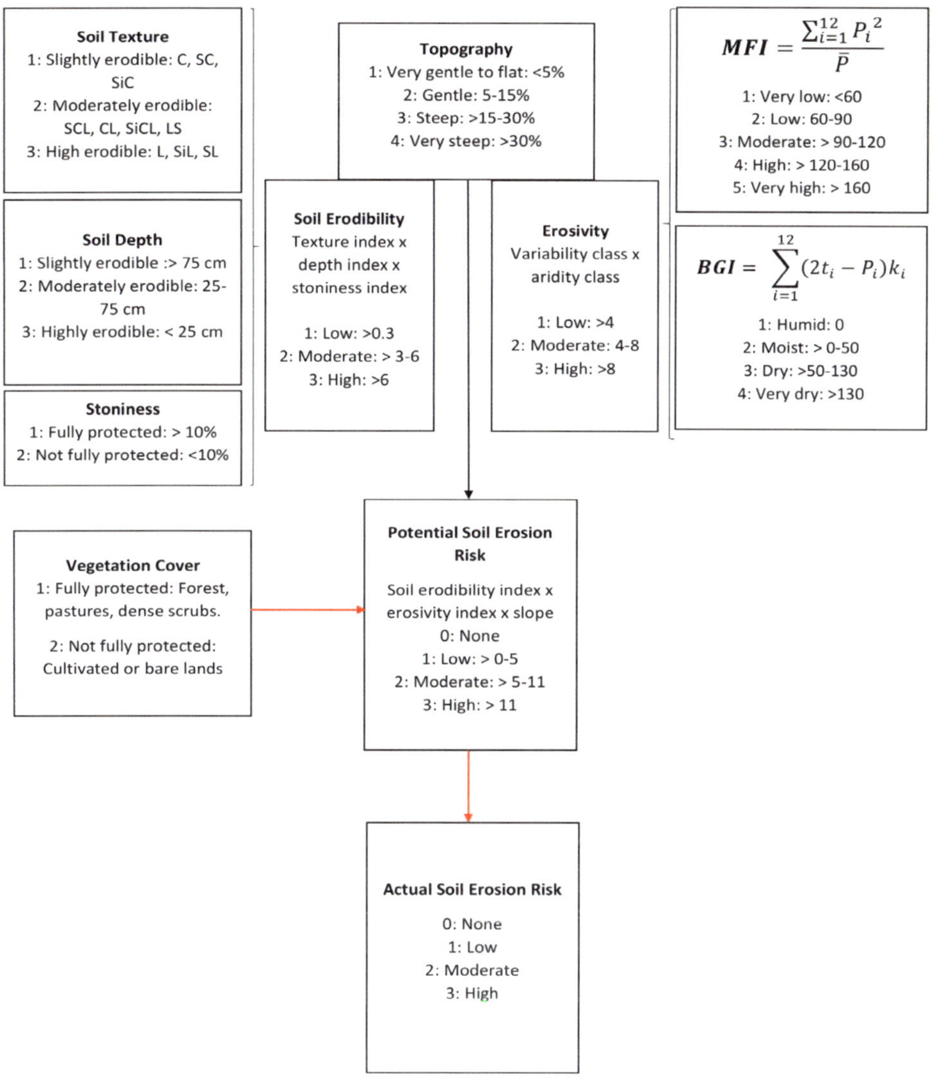

Figure 2.
CORINE model methodology, adapted and modified from CORINE (1992); C: clay, S: sand, Si: silt, L: loam, P_i: total precipitation in month i, Pi: mean annual total precipitation, t_i: mean temperature for the month i, and k_i defined as the proportion of the month in which $2t_i - P_i > 0$.

degraded lands. At the quantitative scale, by computing erosion risks at the current state (reflected by current land occupation) versus LDN state, the quantitative link between the LDN concept and soil erosion by modification of erosion risks after LDN implementation was revealed. For the Claise basin, by means of alternative vegetation cover simulation, the role of ponds on erosion risks was highlighted by revealing changes induced in the shades of their absence or drying.

3.2 Input data and database description

Data availability and quality are one of the main governing factors for any modeling study. The main reason behind the choice of the CORINE model is the data-scarcity state of Nahr Ibrahim where several input data for physical modeling are either lacking or insufficient. Therefore, with respect to the data requirements of the CORINE model, **Table 2** presents the input data for each study area.

3.3 General workflow: a dual approach between current and simulated conditions

The methodological workflow for this study consists of a two-fold approach:

1. Establishment of erosion maps for both study areas under current land occupation settings in order to establish a comparative framework for revealing differences and inferring their sources.

2. Establishment of alternative land use and land cover (LU/LC) scenarios for comparison with current settings: for Nahr Ibrahim based on the LDN concept, and for the Claise basin by means of alternative scenario testing by simulation of pond drying (empty ponds). Alternative simulations are carried out in order to prospect the potential of LDN through land use planning to reduce soil erosion for Nahr Ibrahim, and for determining the pond presence/absence effect in the Claise.

For the LDN approach, the established LU/LC map was intersected under GIS environment with the Lebanese national land capability classification map [95] and the national organic C maps [96] clipped to the Nahr Ibrahim basin in analogy to the LDN indicators. The integration of land capability classification is performed given its importance as an indicator for better use of land, optimization of current LU/LC and for providing insights for future land planning [97, 98]. This step allows a relatively simple yet meaningful tool for land owners and decision-makers for revealing sustainability distribution [66], thus addressing the LDN challenges of land stewardship, and implementing integrated planning approaches for sustainable use of the land and soil resources. After establishing the proposed LDN scenario, based on the concept's response strategy, the LDN-based LU/LC map was used as an alternative input to the CORINE model to compare erosion patterns with those reflecting current conditions in order to reveal LDN's effect on soil erosion in analogy to the work done by Al Sayah et al. [20].

For the Claise basin, study of SAFRAN records for the period 1970–2018, through trend analysis, revealed decreasing precipitations coupled to increases in temperature. Therefore, an alternative scenario assuming that ponds were to be dried was established and inputted again to the CORINE model for comparison with current conditions.

Data	Claise basin	Nahr Ibrahim basin
Land use and land cover maps (classified according to the CORINE classification)	Digitized from ortho-rectified aerial photography 2014, at 0.50 m resolution (R. Nedjai) and verified with ancillary CORINE land use/land cover maps	Digitized from SPOT (2018, 1.5 m) satellite imagery and verified on field—National Council for Scientific Research—Remote Sensing Center
Soil maps	Harmonized World Soil Database [72]	Soil map of Lebanon 1:50000 [82]
DEM	25 m raster; source: Institut Géographique National (IGN) - France	10 m raster, National Council for Scientific Research—Remote Sensing Center
Weather data	Système d'Analyse Fournissant des Renseignements Adaptés à la Nivologie (SAFRAN) model [94]	Lebanese Agricultural Research Institute's Akkoura Weather Station

Table 2.
Input data for the CORINE model and source.

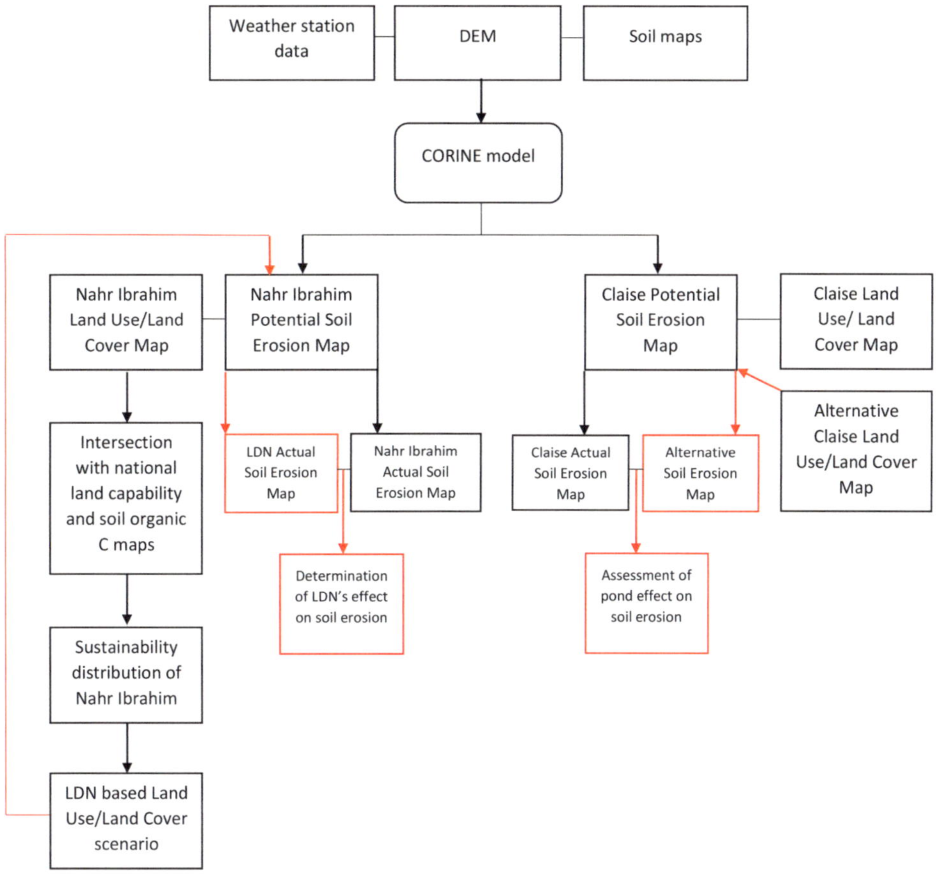

Figure 3.
Methodological workflow.

The produced actual soil erosion map for Nahr Ibrahim was validated on field after a storm event, while for the Claise basin, the actual soil erosion map was validated with ancillary soil erosion maps. **Figure 3** presents the adapted workflow.

4. Results and discussion: comparative analysis of the CORINE's model components for both basins

In this section, a detailed comparison between the two study areas in terms of soil erodibility, erosivity, topography and vegetation cover is first presented. As a second step, the alternative scenarios for both study areas and a comparison with the current conditions for revealing change effects are explained.

4.1 Soil erodibility and pedologic structure of the study areas

With reference to the pedological composition of the study areas, the Claise basin possesses six soil types: Luvisols, Podzols, Leptosols, Cambisols, Fluvisols and Arenosols. On the other hand, the Nahr Ibrahim basin possesses 11 soil types: Leptosols, Andosols, Regosols, Anthrosols, Arenosols, Luvisols, cliffs, Cambisols, Gleysols and Fluvisols. **Tables 3** and **4** present a pedological comparison of the study areas in terms of composition and texture.

With respect to the components of soil erodibility, soil texture in the Claise basin was found to be 68.5% loam, 28.8% loamy sand, 2.5% of clay and the remainder percentage is made of sand. The Nahr Ibrahim basin on the other hand, as function of its more diverse pedological composition, was found to possess more textural classes. These are in descending order of spatial coverage: clay (32.5%), sandy clay loam (26.6%), loamy sand (14.4%), clay loam (10.9%), loam, sandy loam, silty clay loam, silt loam and silty clay. With respect to the CORINE textural classification, the Claise basin mostly corresponds to the highly and moderately erodible texture classes, while Nahr Ibrahim mainly corresponds to the slightly erodible classes. Therefore, in terms of soil texture, the Nahr Ibrahim basin is more erosion resistant than the Claise.

Regarding soil depth, the Claise basin fits to the slightly erodible class with more than 90% of its soils corresponding to the deep (>75 cm) category and the moderately erodible class for its remainder 10%. On the other hand, the Nahr Ibrahim basin presents less than 20% of deep soil classes and more than 40% of shallow depths. Therefore, in terms of soil depth, the Claise basin soils are more resistant to erosion than those of Nahr Ibrahim.

The stone cover of the Claise basin, however, dominantly corresponds to the not fully protected class, while most of the Nahr Ibrahim basin corresponds to the fully protected class, thus giving it a more or less protective stone cover. Globally, the pedological setting of the Nahr Ibrahim basin was found to be more erosion resistant than the Claise.

4.2 Erosivity under different climatic contexts

Since erosivity depends on rainfall, a comparison between the climatic contexts of both study area is presented in **Tables 5** and **6**. As seen, no dry months exist in the Claise and rainfall is much more pronounced than in Nahr Ibrahim. This is observed particularly during summer since rainfall is at its lowest in Nahr Ibrahim as opposed to the Claise where it reaches its maximal values.

At this point, it is important to account for the differences in the climatic settings of both basins, where the Claise corresponds to the degraded oceanic climate, while Nahr Ibrahim is of the Mediterranean type. Therefore, a greater rainfall variability and more prolonged aridity periods are expected for the Nahr Ibrahim, which are characteristic of the Mediterranean climate. This speculation was verified by the Modified Fournier Index (MFI) which was found to be 217 for Nahr Ibrahim (corresponding to the very high erodibility class indexed as 5) and 80 (very low, class 1) for the Claise basin. On the other hand, the Bagnouls-Gaussen aridity index (BGI) further revealed differences between the study areas, where Nahr Ibrahim's BGI is

Claise soil classes	Area (km^2)	Percentage (%)
Calcaric Cambisols	17.38	2.49
Calcaric Fluvisols	4.79	0.69
Cambic Podzols	195.95	28.07
Gleyic Luvisols	439.70	62.99
Luvic Arenosols	2.43	0.35
Rendzic Leptosols	37.83	5.42

Table 3.
Pedological composition of the Claise basin.

Nahr Ibrahim soil classes	Area (km^2)	Percentage (%)
Soil Associations	154.05	50.52
Areno-Eutric Leptosols	16.33	5.36
Calcaric Fluvisols	0.92	0.30
Calcaric Leptosols	1.43	0.47
Calcaric Regosols	3.31	1.08
Calcaro-Hortic Anthrosols	8.18	2.68
Calcaro-Mollic Leptosols	25.29	8.29
Endogleyic Anthrosols	0.25	0.08
Endoskeletic Regosols	6.40	2.10
Eutric Arenosols	0.16	0.05
Eutric Cambisols	1.34	0.44
Eutric Fluvisols	0.46	0.15
Eutric Leptosols	51.78	16.98
Gleyic Leptosols	3.05	1.00
Haplic Arenosols	7.36	2.42
Haplic Luvisols	0.18	0.06
Hypoluvic Arenosols	0.72	0.24
Leptic Andosols	13.34	4.37
Leptic Luvisols	4.09	1.34
Luvic Calcisols	0.02	0.01
Mollic Andosols	0.18	0.06
Mollic Gleysols	1.05	0.35
Rendzic Leptosols	4.82	1.58
Skeletic Regosols	0.16	0.05
Vertic Cambisols	0.05	0.02

Table 4.
Pedological composition of the Nahr Ibrahim basin.

49 (corresponding to the moist class 2) and it is 0 for the Claise corresponding to a humid area with respect to the CORINE BGI classification. In analogy to CORINE's erosivity formula, the Claise basin has an erosivity factor of 1, while for Nahr Ibrahim, the erosivity index is 10. Despite the much more pronounced rainfall in the Claise, the even precipitation distribution in the region resulted in a reduction of climate-induced soil erosion [99] as opposed to Nahr Ibrahim, signifying higher climate-induced erosion risks.

4.3 Effect of topography: a contrast between a mountainous and a flat basin

Table 7 presents the slopes of both study areas with respect to the CORINE's model classification.

Topography is one of the most pronounced differences between the study areas, due to differences in the topographic and orographic composition since the Nahr Ibrahim basin presents a Mediterranean mountainous basin. Accordingly, computing the slope from the DEM rasters of each study area using the slope

Claise month	Jan	Feb	Mar	Apr	May	Jun	Jul	Aug	Sep	Oct	Nov	Dec
Average temp. (°C)	4.7	5.2	7.8	10.4	14.3	17.9	20.1	19.7	16.1	11.8	7.7	4.8
Precipitation (mm)	416	677	1106	1526	1782	2045	2108	1826	1356	818	482	361

Table 5.
Average temperature and precipitation for the Claise basin (1970–2018).

Nahr Ibrahim month	Jan	Feb	Mar	Apr	May	Jun	Jul	Aug	Sep	Oct	Nov	Dec
Average temp. (°C)	4.5	6.1	8.	11.7	15.	18.3	20.6	21	18.4	15.3	10.5	6.8
Precipitation (mm)	181	125	127	56	33	6	0.16	1	19	55	106	197

Table 6.
Average temperature and precipitation for the Nahr Ibrahim basin (2009–2018).

tool of ArcGIS, the Claise basin corresponds entirely to the flat topography class in contrast to the 85% dominance of steep classes in the Nahr Ibrahim basin. For that reason, a significant difference in erosion patterns is expected given the very pronounced role of slope on erosion risks [100], particularly in the Nahr Ibrahim basin, where its slopes, as reported in Ref. [12, 62], were the main reasons behind its high rates of erosion and landslide occurrences as opposed to the predominantly flat Claise basin.

4.4 Vegetation cover: a pronounced difference between a natural and an anthropogenically managed basin

Given its integral role as the most crucial element for erosion risk assessment in the CORINE erosion model, a particular focus is given to the vegetation cover under a setting of natural versus managed basin. This difference is particularly observed when comparing the land use and land cover settings of both study areas. The 707 km^2 Claise basin displays a homogeneous distribution of 21 land occupation classes throughout its area (**Table 8**), while the 309 km^2 Nahr Ibrahim basin occupying an area less than half the area of the Claise presents 43 land use/land cover classes (**Table 9**), which is nearly double the categories of the Claise.

Slope class	Claise (%)	Nahr Ibrahim (%)
Very gentle to flat	99.3	2
Gentle	0.7	13
Steep	—	28
Very steep	—	57

Table 7.
Slope distribution in the study areas.

Claise land occupation	Area (km^2)	Percentage (%)
Agricultural areas	0.15	0.02
Clear broad-leaved forest	2.21	0.31
Clear mixed forest	0.90	0.13
Coniferous forest	26.94	3.70
Dense broad-leaved forest	163.58	23.14
Dense mixed forest	27.00	3.82
Field crops in medium to large terraces	19.14	2.71
Fruit trees	0.20	0.03
Grassland	207.04	29.28
Inland marshes	4.01	0.57
Low-density urban tissue	3.24	0.46
Medium-density urban tissue	1.76	0.25
Mineral extraction site	0.10	0.02
Non-irrigated field crops	151.02	21.36
Pond	79.47	11.24
River	0.55	0.08
Scrubland	2.80	0.40
Scrubland with some bigger dispersed trees	15.36	2.17
Urban expansion site	0.01	0.00
Urban sprawl on clear wooded lands	0.01	0.00
Urban sprawl on field crops	1.01	0.14
Urban sprawl on grassland	1.20	0.17

Table 8.
Distribution of the Claise's land occupation classes.

The land occupation setting of both study areas not only reveals a significant difference between two contexts, but also highlights the effect of management strategies on the studied process. With reference to **Figure 1** and by grouping LU/LC classes into urban/unproductive, agricultural and vegetated (grass, scrublands and forests) areas, a 1.05, 24.07 and 63.01% distribution is observed in the Claise basin, while a 62, 10.27 and 27.73% distribution of the listed class is seen in Nahr Ibrahim. In the Claise, the remainder 11.87% corresponds to water bodies (the Claise River and ponds). Accordingly, with respect to the CORINE erosion model classification, the Claise basin's land occupation pattern corresponds to a 63.58% protected cover and 25.12% not fully protected, while the Nahr Ibrahim basin shows a 29% fully protected cover and a 71% not fully protected. At this point, pronounced differences of topography, climate and vegetation cover are expected to be translated in the erosion maps.

4.5 Actual soil erosion risk maps: a result of contrasting pedological, climatological, topographic and vegetation cover factors

After establishment of the potential soil erosion risk maps in analogy to **Figure 2**, land use and land cover maps of the study areas were intersected to yield the actual soil erosion risk maps of the studied areas (**Figure 4**).

Nahr Ibrahim land occupation	Area (km^2)	Percentage (%)
Medium-density urban tissue	1.7	0.55
Low-density urban tissue	4.16	1.35
Urban expansion sites	0.61	0.2
Industrial or commercial zone	0.31	0.1
Mineral extraction sites	3.18	0.96
Diverse equipment	0.04	0.01
Tourist resorts	0.02	0.01
Field crops in small fields/terraces	4.01	1.02
Urban sprawl on field crops	0.05	0.02
Olives	0.6	0.19
Fruit trees	24.08	7.79
Citrus trees	0.02	0.01
Banana	0.04	0.01
Urban sprawl on permanent crops	2.34	0.76
Greenhouses	0.62	0.2
Dense pine forests	4.42	1.43
Dense oak forests	6.03	1.95
Dense cypress forests	0.08	0.03
Dense juniper forests	0.3	0.1
Dense mixed forests	34.38	11.13
Urban sprawl on dense wooded lands	1.05	0.34
Clear pine forests	1.23	0.4
Clear cypress forests	0.04	0.01
Clear oak forests	9.2	2.98
Clear mixed wooded lands	8.03	2.6
Clear fir forests	0.45	0.15
Clear juniper forests	5.75	1.86
Other type of clear forests	0.05	0.02
Scrublands	2.48	0.8
Scrublands with some bigger dispersed trees	6.34	2.05
Urban sprawl on scrublands	0.0298	0.01
Hill lakes	0.15	0.05
Sand beach	0.03	0.01
Unproductive areas	181.43	58.72
Burnt areas	0.1141	0.04
Abandoned agricultural land	0.74	0.24
Grasslands	5.86	1.9

Table 9.
Distribution of the Nahr Ibrahim's land occupation classes.

The produced maps were in turn verified by field campaigns for the Nahr Ibrahim basin and by cross-validation with ancillary erosion risk maps for the Claise basin. Both maps showed adequate representativity and accuracy in the validation stage. As seen from **Figure 4**, significant differences were observed between the two basins, therefore allowing us to infer several points:

I. Through graphical comparison, the distribution of erosion risks in the two basins is clearly contrasted. The dominance of high erosion risk zones in the Nahr Ibrahim basin is opposed by the prevalence of low erosion risks in the Claise. In the latter, low erosion risks account for 65.6%, moderate risks account for 21.68%, while high erosion risks account for 0.18%. In contrast, the zonal distribution in the Nahr Ibrahim basin is 4% for low risk, 39.5% for moderate risks and 56.42% for high erosion risk zones.

II. The significant difference of erosion patterns between the study areas can be mainly attributed to Nahr Ibrahim's topographic complexity, significant slope steepness, heterogeneous pedological context, dense hydrographic network [31] and its vegetation cover which possesses the most important effect on the CORINE model. Given its status as the only human-controllable input factor, the effect of land management induced by the type of land occupation is also highlighted [101], since a natural setting basin corresponding to a well-managed natural park shows low erosion risks, while a randomly managed basin presents significant erosion levels.

III. In the Claise basin, a no erosion zone is graphically noticed. The latter corresponds to the pond dense zone. At the individual scale, ponds are known for trapping incoming water, increasing its concentration time, decreasing runoff and retaining water, soil and debris by settling, thus trapping eroding soils [102]. At the scale of the Claise, the individual pond effect is much more amplified given the presence of ponds in such large numbers (2179) in a connected matrix, thus increasingly trapping soil/sediment in a collective manner. Their presence as a land occupation class capable of trapping soil and water gives them the role of a protective cover from which soil loss cannot occur, therefore leading to a "no erosion" zone. The collective effect

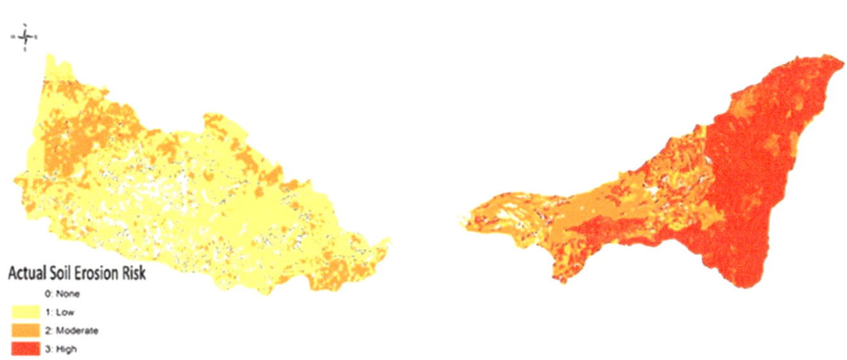

Figure 4.
Actual soil erosion risk for the study areas under current land occupation conditions.

of aggregated ponds as a result of their setting as a conceptual large surface was discussed by Downing [59]; he reports that, as a result of large numbers in an interlocking setting, individual retention capacities and trapping processes are amplified to rates even greater than those of larger water bodies, such as lakes, making these ponds very effective in the process of soil erosion.

IV. Within the Claise basin, moderate erosion risk zones are observed in the agricultural areas. These observations are concurrent with those of Verheijen et al. [8] who report, despite the similarity of the pedological context along with the topographic factor and climatic conditions, soil erosion is ultimately influenced by the vegetation cover and particularly by the presence of agricultural classes (crops) that have been attributed to the highest erosion rates in Europe [4, 103]. Accordingly, despite the homogeneity of the Claise basin and its natural state, agricultural parcels are seen to have higher erosion risks than their surroundings. This further solidifies the role of human-induced LU/LC management in affecting natural processes even within a natural setting.

V. The alarming erosion risk map of Nahr Ibrahim, not only provides an informative tool for erosion, but also highlights the need for intervention, since the basin is severely subjected to soil loss and consequent land degradation. By pin-pointing zones of different erosion risks, an insight toward a priority-based land use planning, targeting zones of higher threats, is achieved. Therefore, in the case of Nahr Ibrahim, soil erosion mapping revealed the spatial distribution of erosion risks as a first step, and served as a land planning decision-oriented tool by pin-pointing zones at high risks as a second step. Through this dual insight provided from the integration of erosion maps, a holistic approach toward land degradation mapping was achieved. Consequently, a proper understanding regarding the types of foreseen soil conservation measures and optimal land occupation classes [104] is made possible, which reiterates the importance of the integration of soil erosion into soil conservation planning [105] and land degradation mapping.

4.6 Alternative simulations for comparison

Analyzing trends obtained from SAFRAN database and applied to the Claise revealed a decreasing trend of precipitation and increasing trend of temperatures. Given the evaporative regime of ponds, an alternative scenario simulating the absence of ponds was obtained. The latter was input, as the alternative vegetation cover, into the CORINE model for comparison with the current condition erosion map in order to determine the impact of pond presence/absence.

For the Nahr Ibrahim basin, the CORINE erosion map provided a tool for land degradation mapping. In analogy to the LDN concept at the scale of soil loss, the land use/land cover, actual soil erosion, national land capability classification and organic C map were intersected to reveal sustainability distribution. The latter was determined following the methodology for sustainability mapping in Al Sayah et al. (2019a) where the adequacy or inadequacy of the already present LU/LC distribution over the different land capability groups (I–IV representing the arable lands and an additional group V combining the USDA's groups V–VIII) allowed the categorization into sustainable and non-sustainable development zones. **Figure 5** shows sustainability distribution in the Nahr Ibrahim basin.

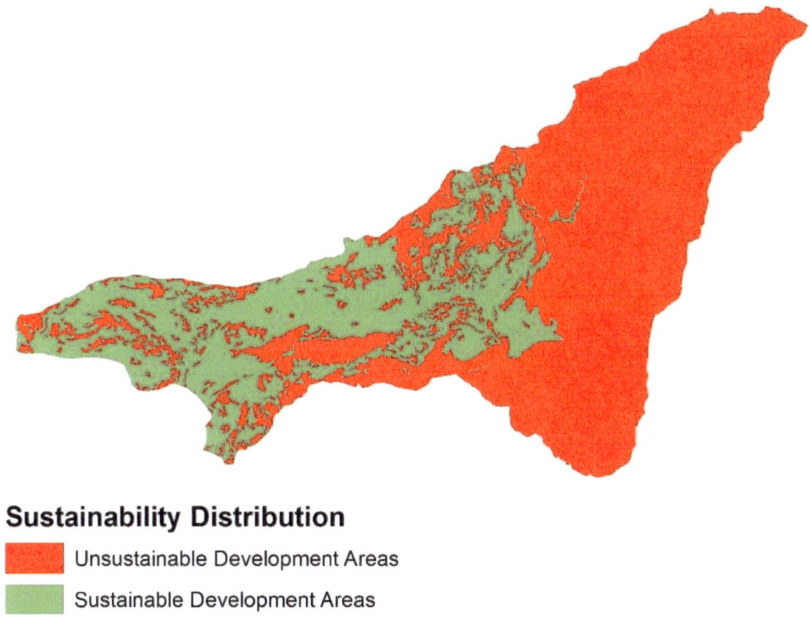

Figure 5.
Sustainability distribution of the Nahr Ibrahim basin obtained from intersecting land use and land cover maps with land capability classification layers.

Noticeably, the prevalence of unsustainable development areas is apparent; these account for 66% of the study area [20]. By optimization of land use and land cover categories covering the soil classes IV and V (19.35% of the basin), an alternative LDN-based scenario was obtained by increasing natural cover (grass, scrublands and forests) over these soils.

By re-using the two alternative vegetation cover scenarios in the CORINE model, **Table 10** was obtained.

Table 10 shows significant shifts of erosion patterns; for Nahr Ibrahim, high erosion risks decreased by 13.9%, low and moderate risks increased by 3 and 10.8%, respectively [20], while for the Claise basin, the opposite was observed with decreases in the no and low erosion risks as compared to increases in the moderate and high erosion risk categories. Thus, the contribution of LDN in reducing erosion highlights the importance of land planning and the effect of management on soil erosion, confining the LDN concept as an effective counter-erosion measure. For the Claise basin, changes in erosion patterns also reveal the importance of ponds as efficient counter-erosion structures that can be used to control areas of significant runoff and excessive erosion.

Erosion risk	Nahr Ibrahim current (%)	Nahr Ibrahim LDN (%)	Claise current (%)	Claise simulation (%)
None	0	0	12.48	1.12
Low	4	7.1	65.66	0.52
Moderate	39.5	50.4	21.68	76.8
High	56.47	42.54	0.18	21.56

Table 10.
Erosion risks of the study areas under current and simulated conditions.

5. Conclusions

As a first step, a simple data demanding CORINE model was used to assess erosion risks of two different geographical settings represented by the Claise and Nahr Ibrahim basins. Several pronounced differences between the two settings were observed as a result of a completely different natural setting and opposing land cover/management practices. A number of conclusions may be drawn from this study; these are listed under fundamental and contextual settings.

1. Fundamentally, despite the abundance of several erosion models, given the data-scarcity of Nahr Ibrahim and for the purpose of comparison between the two study areas, the relatively simple data demanding CORINE model was used. As a first step, the CORINE erosion model proved to be a robust tool for evaluation of the spatial distribution of erosion risks despite its empirical nature where CORINE established maps have shown sufficient accuracy when verified on field and crossed with ancillary maps.

2. In addition to erosion assessment, the CORINE model serves as a proficient tool for land occupation and land management adequacy assessment given its vegetation cover input that reveals the actual erosion risk settings of basins under current conditions. This statement was justified by intersecting land use and land cover maps with the actual soil erosion risk map in Nahr Ibrahim revealing the extent of mismanagement as function of inadequate allocation. In addition, by highlighting zones of high risks, an insight towards prioritized treatment measures is obtained. Moreover, by revealing zones of different risk levels, the CORINE model provides insight for land use planning, thus promoting optimal land occupation allocation. Further, by changing the vegetation cover input as the human-controllable factor and stabilizing all other components, the CORINE model serves also as a tool for alternative scenario assessment by revealing changes of erosion patterns under different scenarios when compared with the current baseline conditions of the studied area, thus revealing the needed steps to follow in terms of land planning or soil and water conservation measures.

3. In Mediterranean settings such as the Nahr Ibrahim basin, the CORINE model can provide a starting point for combatting land degradation, thus filling gaps of LDN application in the Mediterranean basin by contributing to land degradation mapping, integration of site-specific land degradation drivers and promoting sustainable land use planning [64, 106, 107].

Contextually, and by comparing both study areas, several aspects can be pointed out. Despite differences in the geographical setting, the impact of adequate versus random land use planning can be first concluded. This statement is particularly justified in the Claise basin, where despite its challenging pedological settings in terms of weak structure and cover, low and moderate erosion risks are prevalent due to its natural setting that provides the basin a protective cover against erosion. Further, due to the presence of ponds in large numbers, an amplified counter-erosion effect is observed. Their role was solidified by fixing erosivity, erodibility and topographic factors of the model and inputting an alternative scenario with dredged ponds. By comparison with the current actual soil erosion risk map, not only a shift in local erosion risks was observed, but also a complete shift within the basin was shown, thus confining the low erosion state of the Claise to its natural and pond cover and further indicating the efficiency of projecting ponds as an effective counter-erosion measure in basins with high erosion risks such as the Nahr Ibrahim basin.

When comparing the erosive setting of the Claise with Nahr Ibrahim, significant differences were observed namely in high erosion risk zones. This, in turn, is attributed to the climatic, topographic and vegetation cover factors of Nahr Ibrahim where increased climate-induced erosion combined with the very steep slope and anthropogenically induced erosion from alteration of the vegetation cover is prevalent. Under current conditions, the land occupation pattern of Nahr Ibrahim was shown to be unsustainable in terms of distribution above lands of different capabilities and distribution along high erosion risk areas. The most striking difference between the two basins is that the Nahr Ibrahim accounts for nearly double the number of land occupation classes in the Claise basin for an area less than its half. Further, the unequal repartition of land use/land cover classes in the Nahr Ibrahim basin caused a gradient of soil erosion risk patterns, consisting mainly of high erosion risks in its upper section and moderate to low risks in its middle and lower parts.

Despite its (the Nahr Ibrahim basin's) pedological and topographic settings, when vegetation cover was optimized through the application of the LDN concept, erosion risks significantly shifted. This is attributed to its highly erosive state and to its land occupation and management pattern in contrast to the well-controlled Claise basin. Conversely, the use of LDN as a basis for land planning and the use of land planning for implementation of the LDN concept not only allowed sustainability restoration but also proved to be an effective counter-erosion tool given its effect on decreasing high erosion risks and increasing low and moderate ones. The coupling of the CORINE erosion model and LDN concept can play a role in decision-making regarding land use planning, thus highlighting the importance of their implementation at the scale of the Mediterranean landscape. However, a basin like Nahr Ibrahim cannot be converted into a setting similar to the Claise, but a balanced land use plan accounting for the trade-off between natural resources and urban expansion may be the solution for restoring the Nahr Ibrahim landscape.

Finally, through a simple methodological approach, this work can be listed as a response to the European framework for the Thematic Strategy on Soil Protection, recommendations of the DCE for revealing the role of hydromorphological alternating structures on erosion patterns in basins and UNCCD's recommendations for implementation of the LDN concept. Despite the differences between the Thematic Strategy on Soil Protection, DCE and LDN concepts, the common effect of land occupation within these frameworks can be used as a platform to study the extent of anthropogenic influence at the basin scale in an attempt to promote sustainable development and to integrate soil erosion into land planning.

Acknowledgements

This research is part of a PhD thesis funded by the National Council of Scientific Research-Lebanon (CNRS-L), Agence Universitaire de la Francophonie (AUF), Lebanon and the Lebanese University. It is also part of the Dynétangs project funded by the Centre-Val-de-Loire region. The Authors would like to extend their gratitude to the editors of the book and to the Brenne Regional Natural Park for their help in weather station maintenance, and to Dr. Ihab Jomaa of the Lebanese Agricultural Research Institute (LARI) for providing weather station data.

Conflict of interest

The authors declare no conflict of interest.

Author details

Mario J. Al Sayah[1,2,3*], Rachid Nedjai[3], Chadi Abdallah[1], Michel Khouri[2], Talal Darwish[1] and François Pinet[4]

1 National Council for Scientific Research, Remote Sensing Center, Beirut, Lebanon

2 Centre de Recherches en Sciences et Ingénierie, Lebanese University Faculty of Engineering II, Roumieh, Lebanon

3 Centre d'Études et de Développement des Territoires et de l'Environnement, Université d'Orléans, Orléans, France

4 Brenne Regional Natural Park, Rosnay, France

*Address all correspondence to: mario_sayah94@hotmail.com

References

[1] Pimentel D. Soil erosion: A food and environmental threat. Environment, Development and Sustainability. 2006;**8**:119-137

[2] Lal R. Soil erosion and the global carbon budget. Environment International. 2003;**29**(4):437-450

[3] Keesstra SD, Bouma J, Wallinga J, Tittonell P, Smith P, Cerdà A, et al. The significance of soils and soil science towards realization of the United Nations sustainable development goals. The Soil. 2016;**2**:111-128

[4] FAO. International Technical Panel on Soils. Status of the World's Soil Resources. Rome, Italy: FAO; 2015

[5] Zuazo VHD, Pleguezuelo CRR. Soil-erosion and runoff prevention by plant covers. A review. Agronomy for Sustainable Development. 2008;**28**:65-86

[6] Panagos P, Borrelli P, Poesen J, Ballabio C, Lugato E, Meusburger K, et al. The new assessment of soil loss by water erosion in Europe. Environmental Science & Policy. 2015;**54**:438-447

[7] Terranova O, Antronico L, Coscarelli R, Laquinta P. Soil erosion risk scenarios in the Mediterranean environment using RUSLE and GIS: An application model for Calabria (southern Italy). Geomorphology. 2009;**112**:228-245

[8] Verheijen FGA, Jones RJA, Rickson RJ, Smith CJ. Tolerable versus actual soil erosion rates in Europe. Earth-Science Reviews. 2009;**94**:23-38

[9] El Bagouri IH. Land degradation control in northern Africa. In: Sivakumar NVK, Ndiang'ui N, editors. Climate and Land Degradation. Berlin, Heidelberg: Springer; 2007. pp. 391-340

[10] Darwish T, Faour G, Khawlie M. Assessing soil degradation by landuse-cover change in coastal Lebanon. Lebanese Science Journal. 2004;**5**(1):45-60

[11] Peñuelas J, Sardans J, Filella I, Estiarte M, Llusià J, Ogaya R, et al. Impacts of global change on Mediterranean forests and their services. Forests. 2017;**8**(463):37

[12] Bou Kheir R, Cerdan O, Abdallah C. Regional soil erosion risk mapping in Lebanon. Geomorphology. 2006;**82**(3-4):347-359

[13] Gobin A, Govers G, Jones R, Kirkby M, Kosmas C. Assessment and Reporting on Soil Erosion. Background and Workshop Report. Technical Report No. 84; 2002

[14] Abdelwahab OM, Ricci GF, De Girolamo AM, Gentile F. Modelling soil erosion in a Mediterranean watershed: Comparison between SWAT and AnnAGNPS models. Environmental Research. 2018;**166**: 363-376

[15] Darwish T, Atallah T, Fadel A. Challenges of soil carbon sequestration in the NENA region. The Soil. 2018;**4**:225-235

[16] D'Ostiani LF. Watershed Management: A Key Component of Rural Development in the Mediterranean Region. Rome: FAO; 2004

[17] Verdeil E, Faour G, Velut S, editors. Atlas du Liban les nouveaux défis. Presses de l'Ifpo - CNRS Liban, Beirut; 2016

[18] García-Ruiz JM, Nadal-Romero E, Lana-Renault N, Beguería S. Erosion in Mediterranean landscapes: Changes and future challenges. Geomorphology. 2013;**198**:20-36

[19] Piccarreta M, Capolongo D, Boenzi F, Bentivenga M. Implications of decadal changes in precipitation and land use policy to soil erosion in Basilicata, Italy. Catena. 2006;**65**:138-151

[20] Al Sayah MJ, Abdallah C, Khouri M, Nedjai R, Darwish T. Application of the land degradation neutrality concept in Mediterranean watersheds, a case study of Nahr Ibrahim, Lebanon. Geophysical Research Abstracts. 2019;**21**(2019-3510):1

[21] Herrmann SM, Hutchinson CF. The changing contexts of the desertification debate. Journal of Arid Environments. 2005;**63**(3):538-555

[22] Vogt JV, Safriel U, von Maltitz G, Sokona Y, Zougmore R, Bastin G, et al. Monitoring and assessment of land degradation and desertification: Towards new conceptual and integrated approaches. Land Degradation and Development. 2011;**22**:150-165

[23] Reynolds JF, Smith D, Stafford M, Lambin EF, Turner BL, Mortimore M, et al. Global desertification: Building a science for dryland development. Science. 2007;**316**(5826):847-851

[24] Panagos P, Borelli P. Soil erosion in Europe: Current status, challenges and future developments. All That Soil Erosion: the Global Task to Conserve Our Soil Resources. Korea: Soil Environment Center of Korea; 2017

[25] Grainger A. Is land degradation neutrality feasible in dry areas? Journal of Arid Environments. 2015;**112**(A):14-24

[26] Safriel UN. Status of desertification in the Mediterranean region. In: Rubio JL, Safriel U, Daussa R, Blum W, Pedrazzini F, editors. Water Scarcity, Land Degradation and Desertification in the Mediterranean Region. Dordrecht: Springer, Dordrecht; 2009. pp. 33-73

[27] Pradeep GS, Krishnan MVN, Vijith H. Identification of critical soil erosion prone areas and annual average soil loss in an upland agricultural watershed of Western Ghats, using analytical hierarchy process (AHP) and RUSLE techniques. Arabian Journal of Geosciences. 2015;**8**:3697-3711

[28] Kim J, Ivanov VY, Fatichi S. Soil erosion assessment—Mind the gap. Geophysical Research Letters. 2016;**43**(12):446-454

[29] Paroissien JB, Darboux F, Couturier A, Devilliers B, Mouillot F, Raclot D, et al. A method for modeling the effects of climate and land use changes on erosion and sustainability of soil in a Mediterranean watershed (Languedoc, France). Journal of Environmental Management. 2015;**150**:57-68

[30] Mosbahi M, Benabdallah S, Boussema MR. Assessment of soil erosion risk using SWAT model. Arabian Journal of Geosciences. 2012;**6**(10):4011-4019

[31] Kaffas K, Hrissanthou V. Soil erosion, streambed deposition and streambed erosion—Assessment at the mountainous terrain. Proceedings. 2018;**2**(626):1-8

[32] Pimentel D, Burgess M. Soil erosion threatens food production. Agriculture. 2013;**3**:443-463

[33] Dotterweich M. The history of human-induced soil erosion: Geomorphic legacies, early descriptions and research, and the development of soil conservation—A global synopsis. Geomorphology. 2013;**201**:1-34

[34] Boardman J, Poesen J, Evans R. Socio-economic factors in soil erosion

and conservation. Environmental Science & Policy. 2003;**6**:1-6

[35] Keesstra SD, Geissen V, Mosse K, Piiranen S, Scudiero E, Leistra M, et al. Soil as a filter for groundwater quality. Current Opinion in Environment Sustainability. 2012;**4**:507-516

[36] Alam A. Soil degradation: A challenge to sustainable agriculture. International Journal of Agricultural Science and Research. 2014;**1**(4):50-55

[37] Hediger W. Sustainable farm income in the presence of soil erosion: An agricultural Hartwick rule. Ecological Economics. 2003;**45**:221-236

[38] Jie C, Jing-zhang C, Man-zhi T, Zi-tong G. Soil degradation: A global problem endangering sustainable development. Journal of Geographical Sciences. 2002;**12**(2):243-252

[39] van Rijn LC. Principles of Sediment Transport in Rivers, Estuaries and Coastal Seas. Amsterdam, Netherlands: Aqua Publications; 1993. 690 p

[40] Kaffas K, Hrissanthou V, Sevastas S. Modeling hydromorphological processes in a mountainous basin using a composite mathematical model and ArcSWAT. Catena. 2018;**162**(April):108-129

[41] Ffolliott PF, Brooks KC, Neary DG, Tapia PR, Chevesich PG. Soil Erosion and Sediment Production on Watershed Landscapes: Processes and Control. UNESCO Special Technical Publication No. 32. Montevideo, Uruguay: UNESCO, International Hydrological Programme, Regional Office for Science for Latin American and the Carribbean. 2013. p. 73

[42] Bou Kheir R, Abdallah C, Khawlie M. Assessing soil erosion in Mediterranean karst landscapes of Lebanon using remote sensing and GIS. Engineering Geology. 2008;**99**:239-254

[43] Wischmeier WH, Smith DD, Uhland RE. Evaluation of factors in the soil-loss equation. Agricultural Engineering. 1958;**39**:458-462

[44] Wischmeier W, Smith D. Predicting Rainfall Erosion Losses: A Guide to Conservation Planning. USA: Department of Agriculture Handbook No. 537; 1978

[45] Quinton J. Reducing predictive uncertainty in model simulations: A comparison of two methods using the European soil erosion model (EUROSEM). Catena. 1997;**30**:101-117

[46] Flanagan DC, Gilley JE, Franti TG. Water erosion prediction project (WEPP) development history, model capabilities and future enhancements. American Society of Agricultural and Biological Engineers. 2007;**50**(5):1603-1612

[47] CORINE. CORINE Soil Erosion Risk and Important Land Resources in the Southern Regions of the European Community. Luxembourg: European Environment Agency; 1992

[48] Vertessy RA, Hatton TJ, O'Shaughnessy PJ, Jayasuriya MDA. Predicting water yield from a mountain ash forest catchment using a terrain analysis based catchment model. Journal of Hydrology. 1993;**150**(2-4):665-700

[49] Wilkinson S, Henderson A, Chen Y. SedNet a CSIRO Land and Water client report for the Cooperative Research Centre for Catchment Hydrology [Internet]. 2008. Available from: www.toolkit.net.au/sednet USER

[50] Stolte J, Tesfai M, Øygarden L, Kværnø S, Keizer J, Verheijen F, et al. Soil Threats in Europe: Status, Methods, Drivers and Effects on Ecosystem

Services. Luxembourg: European Commission Joint Research Center; 2016

[51] Abdallah C, Der Sarkissian R, Termos S, Darwich T, Faour G. Agricultural Risk Assessment for Lebanon to Facilitate Contingency & DRR/CCA Planning. Beirut, Lebanon: FAO; 2018

[52] Cerdan O, Govers G, Le Bissonnais Y, van Oost K, Poesen J, Saby N, et al. Rates and spatial variations of soil erosion in Europe: A study based on erosion plot data. Geomorphology. 2010;**122**(1-2):167-177

[53] Vanmaercke M, Matens W, Poesen J, Jankauskas B, Jankauskiene G, Verstraeten G, et al. A comparison of measured catchment sediment yields with measured and predicted hillslope erosion rates in Europe. Journal of Soils and Sediments. 2012;**12**:586-602

[54] Duru U, Arabi M, Wohl EE. Modeling stream flow and sediment yield using the SWAT model: A case study of Ankara River basin, Turkey. Physical Geography. 2017;**39**(3):264-289

[55] Abdallah C. Application of Remote Sensing and Geographical Information System for the Study of Mass Movements in Lebanon. Université Pierre et Marie Curie Paris 6; 2007

[56] Zuazo VHD, Martínez JRF, Tejero IG, Pleguezuelo CRR, Raya AM, Tavira S. Runoff and sediment yield from a small watershed in southeastern Spain (Lanjarón): Implications for water quality. Hydrological Sciences Journal. 2012;**57**(8):1610-1625

[57] Nedjai R, Azaroual A, Chlif K, Bensaid A, Al Sayah M, Ysbaa L. Impact of ponds on local climate: A remote sensing and GIS contribution application to the ponds of Brenne (France). Journal of Earth Science and Climatic Change. 2018;**9**(12):10

[58] Verstraeten G, Poesen J. Estimating trap efficiency of small reservoirs and ponds: Methods and implications for the assessment of sediment yield. Progress in Physical Geography. 2000;**24**(2):219-251. DOI: 10.1177/030913330002400204

[59] Downing JA. Emerging global role of small lakes and ponds: Little things mean a lot. Limnetica. 2010;**29**(1):9-24

[60] Commission Européenne. La directive-cadre européenne sur l'eau. Union Européenne: Comission Européenne Office des Publications; 2000

[61] Darwish T, Shaban A, Portoghese I, Vurro M, Khadra R, Saqallah S, et al. Inducing water productivity from snow cover for sustainable water Management in Ibrahim River Basin, Lebanon. British Journal of Applied Science & Technology. 2015;**5**(3): 233-243 Available from: http://www.sciencedomain.org/abstract.php?iid=760&id=5&aid=6610

[62] Abdallah C, Faour G. Landslide hazard mapping of Ibrahim River basin, Lebanon. Natural Hazards. 2016;**85**(1):237-266

[63] Zhu M. Soil erosion risk assessment with CORINE model: Case study in the Danjiangkou reservoir region, China. Stochastic Environmental Research and Risk Assessment. 2012;**26**(6):813-822

[64] Briassoulis H. Combating land degradation and desertification: The land-use planning quandary. Land. 2019;**8**(27):1-26

[65] Orr BJ, Cowie A, Castillo Sanchez VM, Chasek P, Crossman ND, Erlewein A, et al. Scientific Conceptual Framework for Land Degradation Neutrality. A Report of the Science-Policy Interface. Bonn, Germany; 2017

[66] Al Sayah MJ, Abdallah C, Khouri M, Nedjai R, Darwich T. Application of

the LDN concept for quantification of the impact of land use and land cover changes on Mediterranean watersheds - Al Awali basin - Lebanon as a case study. Catena. 2019;**176**(May):264-278

[67] Communauté de Communes Loches Sud Touraine. Dossier préalable à la Déclaration d'Intérêt Général (D.I.G), l'Autorisation Environnementale. Installations, ouvrages, travaux et activités; 2017

[68] Guichané R. L'aménagement hydraulique de la Claise tourangelle et de ses affluents du Moyen-Âge à nos jours / Mills on the claise and its tributaries in Indre-et-Loire from the middle ages to modern times. Revue archéologique du Centre de la France. 1993;**32**:109-152

[69] SANDRE. Fiche cours d'eau la Claise (L6-0200); 2012

[70] Bouscasse H, Defrance P, Amand B, Grandmougi B, Strosser P, Beley Y. Amélioration des connaissances sur les fonctions et usages des zones humides: évaluation économique sur des sites tests, le cas des étangs de la Grande Brenne; 2011

[71] Joly D, Brossard T, Cardot H, Cavailhes J, Hilal M, Wavresky P. Les types de climats en France, une construction spatiale - Types of climates on continental France, a spatial construction. Cybergéo European Journal of Geography. 2010;**501**:1-23. Available from: http://prodinra.inra.fr/ft?id=%7BDDCC3C34-7355-486A-A7D2-E2509D33DC4F%7D%5Cnhttp://cybergeo.revues.org/index23155.html

[72] Fischer G, Nachtergaele F, Prieler S, van Velthuizen HT, Verelst L, Wiberg D. Global Agro-Ecological Zones Assessment for Agriculture (GAEZ 2008). Rome, Italy: FAO; 2008

[73] Fischer G, Nachtergaele F, Prieler S, van Velthuizen HT, Verelst L, Wiberg D. Global Agro-Ecological Zones Assessment for Agriculture (GAEZ 2008). Rome, Italy: FAO; 2008

[74] Barrier P, Gagnaison C. Notice explicative de la feuille Le Blanc à 1/50 000. BRGM Orléans; 2005

[75] Benarrous R. La Grande Brenne aux périodes préindustrielles (Indre). Contribution à l'histoire des paysages, des étangs et des relations sociétés/milieux dans une zone humide continentale. Approches historique, archéologique et paléoenvironnementale. France: Université de Paris I - Panthéon Sorbonne; 2009

[76] Chalmeton P, Géniteau F, Cherbonnet C, Bourdais A, Leclerc C. Diagnostic local de santé du Parc naturel régional de la Brenne en vue de l'élaboration du Contrat Local de Santé. Brenne; 2015

[77] Chalmeton P, Géniteau F, Cherbonnet C, Bourdais A, Leclerc C. Diagnostic local de santé du Parc naturel régional de la Brenne en vue de l'élaboration du Contrat Local de Santé. Parc Naturel Regional de la Brenne, Agence Régionale de Santé Centre, ORS Centre; 2015

[78] MoE/UNDP. Strategic Environmental Assessment of Lebanon's Renewable Energy Sector. Beirut, Lebanon: MoE and UNDP; 2014

[79] MoE/UNDP. Strategic Environmental Assessment of Lebanon's Renewable Energy Sector. Lebanon: MoE and UNDP Beirut; 2014e

[80] Leung KY. One-Dimensional Model of Fecal Coliform in Nahr Ibrahim River (Lebanon). USA: University of Illinois, Massachusetts Institute of Technology; 2001

[81] Assaker A. Hydrologie et Biogéochimie du Bassin Versant du

Fleuve Ibrahim: Un Observatoire du Fonctionnement de la Zone Critique au Liban [Internet]. France: Université de Toulouse; France. 2016. Available from: https://oatao.univtoulouse. fr/15610/1/ASSAKER.pdf

[82] Darwish T, Khawlie M, Jomaa I, Abou Daher M, Awad M, Masri T, et al. Soil Map of Lebanon: 1:50 000. Monograph. Beirut, Lebanon: CNRS, Remote Sensing Center; 2006. 367 p

[83] Dubertret L. Cartes geologiques du Liban à l'echelle 1 :50000; 1955

[84] Hreiche A, Najem W, Bocquillon C. Hydrological impact simulations of climate change on Lebanese coastal rivers. Hydrological Sciences Journal. 2007;**52**(6):1119-1133

[85] Liu BY, Nearing M, Shi PJ, Jia ZW. Slope length effects on soil loss for steep slopes. Soil Science Society of America Journal. 2000;**64**:1759-1763

[86] Gurebiyaw K, Addis HK, Teklay A. Assessment of spatial soil erosion susceptibility based on the CORINE model in the Gumara-Maksegnit watershed, Ethiopia. Journal of Natural Resources and Development. 2018;**8**:38-45

[87] Yuksel A, Gundogan R, Akay AE. Using the remote sensing and GIS technology for erosion risk mapping of Kartalkaya dam watershed in Kahramanmaras, Turkey. Sensors. 2008;**8**:4851-4865

[88] Tayebi M, Tayebi MH, Samenim A. Soil erosion risk assessment using GIS and CORINE model: A case study from western Shiraz, Iran. Archives of Agronomy and Soil Science. 2017;**63**(8):1163-1175

[89] Giordano A. The CORINE project on soil erosion risk and land quality. Land use, land cover and soil sciences. In: Systems Engineering and Management for Sustainable Development. Encyclopedia. Vol. III. Italy: Università di Torino; 2009. p. 27

[90] Wang D, Fu B, Zhao W, Hu H, Wang Y. Multifractal characteristics of soil particle size distribution under different land-use types on the Loess Plateau, China. Catena. 2008;**72**:29-36

[91] Fournier F. Climat et Erosion: La Relation Entre L'Erosion Du Sol Par L'Eau Et Les Précipitations Atmospheriques. Presses Universitaires De France, editor. Paris: Publication de l'Institut d'Études Roumaines; 1960. 201 p

[92] Gaussen H. Bioclimatic Map of Mediterranean Zone. Paris, France; 1963

[93] De Baets S, Poesen J, Knapen A, Barberá GG, Navarro JA. Root characteristics of representative Mediterranean plant species and their erosion-reducing potential during concentrated runoff. Plant and Soil. 2007;**294**:169-183

[94] MétéoFrance. Présentation générale du modèle de surface; 2016

[95] Darwish T, Jomaa I, Awad M, AbouDaher M, Msann J. Inventory and management of Lebanese soils integrating the soil geographical database of Euro-Mediterranean countries. Lebanese Science Journal. 2005;**6**(2):15

[96] Darwish T, Fadel A. Mapping of soil organic carbon stock in the Arab countries to mitigate land degradation. Arabian Journal of Geosciences. 2017;**10**(474):1-11

[97] Abou-Najem S, Palacios-Rodríguez G, Darwish T, Faour G, Kattar S, Clavero Rumbao I, et al. Land capability for agriculture, Hermel District, Lebanon. Journal of Maps. 2019;**15**:1-10

[98] Gashaw T, Tulu T, Argaw M, Worqlul AW. Land capability

classification for planning land uses in the Geleda watershed, Blue Nile Basin, Ethiopia. Modeling Earth Systems and Environment. 2018;**4**(2):489-499

[99] Ahmed SI, Rudra RP, Gharabaghi B, Mackenzie K, Dickinson WT. Within-storm rainfall distribution effect on soil erosion rate. ISRN Soil Science. 2012;**2012**:1-7

[100] Zhang Z, Sheng L, Yang J, Chen XA, Kong L, Wagan B. Effects of land use and slope gradient on soil erosion in a red soil hilly watershed of southern China. Sustainability. 2015;7:14309-14325

[101] Jinren RN, Yingkui KL. Approach to soil erosion assessment in terms of land-use structure changes. Journal of Soil and Water Conservation. 2003;**58**(3):158-169

[102] Ismail T, Othman MA, Fadzil AB, Zainuddin ZM. Deposition of sediments in detention pond. Malaysian Journal of Civil Engineering. 2010;**22**(1):95-118

[103] Panagos P, Standardi G, Borrelli P, Lugato E, Montanarella L, Bosello F. Cost of agricultural productivity loss due to soil erosion in the European Union: From direct cost evaluation approaches to the use of macroeconomic models. Land Degradation and Development. 2018;**29**:471-484

[104] Duan X, Shi X, Li Y, Li R, Fen D. A new method to calculate soil loss tolerance for sustainable soil productivity in farmland. Agronomy for Sustainable Development. 2017;**37**(1):1-13

[105] Dutta S. Soil erosion, sediment yield and sedimentation of reservoir: A review. Modeling Earth Systems and Environment. 2016;**2**(123):1-18

[106] Chasek P, Safriel U, Shikongo S, Fuhrman VF. Operationalizing zero net land degradation: The next stage ininternational efforts to combat desertification? Journal of Arid Environments. 2015;**112**(A):5-13

[107] Wunder S, Kaphengst T, Larsen AF. Implementing land degradation neutrality (SDG 15.3) at National Level: General approach, indicator selection and experiences from Germany. In: Ginzky H, Dooley E, Heuser IL, Kasimbazi E, Markus T, Qin T, editors. International Yearbook of Soil Law and Policy. Germany: Springer; 2017. pp. 191-219

Chapter 5

Spatial Analysis of the Erosive Hazard of Soils and Natural Risks of Reservoir Siltation

Rabii El Gaatib

Abstract

The initial state of several watersheds, in West Africa, is characterized by a socio-ecological vulnerability linked to the water erosion risks. Thus, the Oued Beht watershed (430,728 ha), which is located in Morocco, reveals the extent of impact of soil erosion and water quality degradations. Especially, the consequences of soil loss alter its hydrological behavior in terms of efficiency to produce good water quality and include damages to the functional activities (agricultural and forestry) and structural challenges (lands and dams). This study suggests a methodology, reproducible and generalizable, to assess the water erosion risks. The results show that the erosion process is characterized by the combination of several types of erosion including sheet, rill, and gully. Therefore, the soil erosion is active and visible on more than 3/4 of the Oued Beht watershed, and the spatial analysis evaluates the soil loss which generates a decrease in the storage capacity of El Kansra dam (−3.03 million m^3/year). The erosion risk management is evaluated by combining susceptibility maps with an analysis of potential consequences. Moreover, the interactive mode obtained from this work is based on a statistical autocorrelation approach concerning risk factors in order to delimit the areas requiring priority planning (hot spots).

Keywords: soil erosion, watershed, hot spots, spatial autocorrelation, risk management

1. Introduction

The soil erosion characterizes the majority of Morocco reliefs, and a spectacular expansion of erosion processes reveals more disturbing aspects. Thus, the soils degradation upstream is the origin of siltation phenomenon and decreasing storage capacities of dams with 50 million m^3/year [1]. Particularly, the erosion hazard imposes significant costs on the Moroccan economy by reducing soil productivity and the consequences are manifested by dam siltation downstream.

In this sense, the results obtained in the first phase of this study have shown the importance of erosion in Oued Beht watershed revealing that combined forms are meaningful (sheet, rill, and gully) and many factors, both physical and human, promote erosion risk. Moreover, the human context is generally characterized by high density of rural population [2, 3].

In this perspective, this study provides a roadmap relating to biophysical, hydrological, and socio-economic backgrounds to develop a dynamic methodology that will identify and visualize development scenarios. The specific objectives that are identified include the following:

- Analyzing the biophysical context and highlighting the environmental potentialities and constraints.
- Developing the spatial modeling of soil and water degradation processes with integration of empirical models in a GIS environment to determine the potential soil loss.
- Prescribing the strategic orientations of Master Plan Management to allow the sustainability of the main actions linked to erosion control.
- Defining the action plan to be used in priority areas and identify the biological measures and appropriate soil conservation practices to be implemented in order to mitigate negative effects of erosion hazard.

2. Study area

The Oued Beht watershed is located upstream of El Kansra dam (85 km east from Rabat), which crosses the Central Highlands and the Middle Atlas of Morocco (**Figure 1**). The main stream is Oued Beht, affluent of Sebou river, one of the most important watersheds in the kingdom. Thus, the watershed overlaps the administrative territory erected into 5 provinces and 26 rural communes (**Figure 2**).

The delimitation of the watershed in the geographic information system (GIS) provides a total area of 430,728 ha with an elongated form (**Figure 2**). It owns a developed urban system, occupying a central place in socio-economic

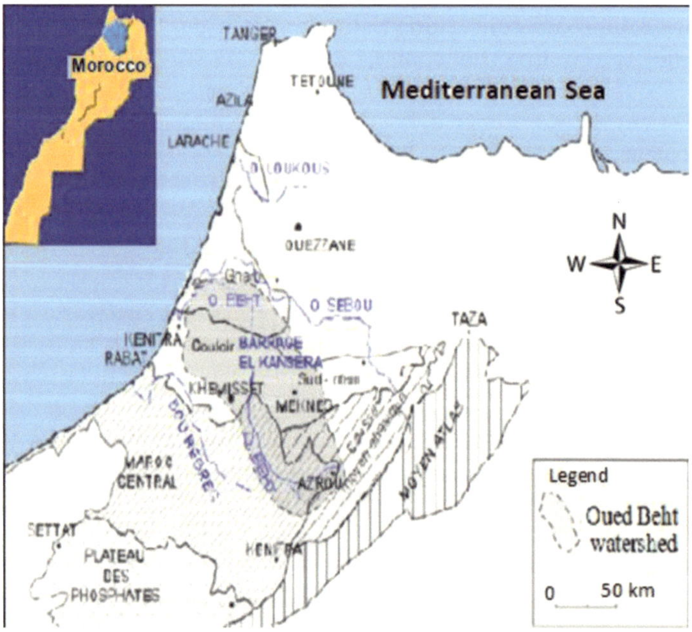

Figure 1.
Localization of the study area inside Morocco country.

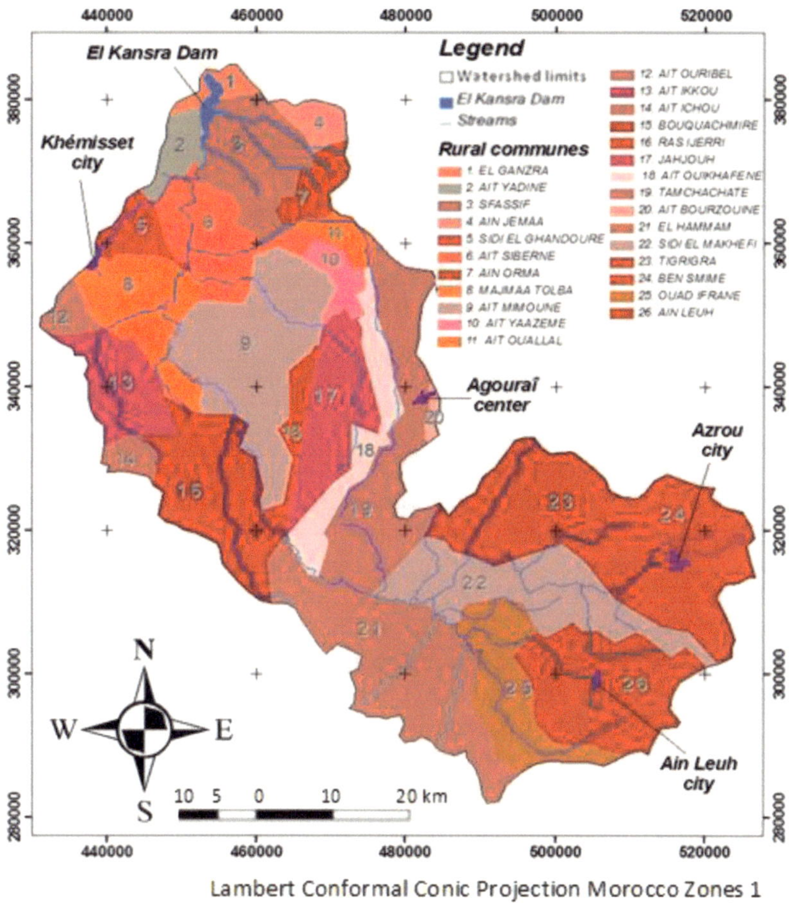

Figure 2.
Distribution map of rural communes in the Oued Beht watershed.

activities; it is Khemisset city (542,000 inhabitants), Azrou (47,540 inhabitants), and urban centers of Agourai and Ain Leuh [4]. Concerning climate context, the watershed presents characteristics of Mediterranean climate with a rainy winter and a dry summer [5].

The coordinates of the map (longitude is x and latitude is y) are described in a map projection. Thus, the cartographic representation of the whole watershed surface on a two-dimensional map (X, Y) is based on the use of the Lambert conformal conic projection. Consequently, the Oued Beht watershed is located between the rectangle designated by the following Lambert coordinates: (X1 = 430,347, Y1 = 282,142) and (X2 = 527,857, Y2 = 383,572).

3. Materials and methods

The adopted methodological framework allows meeting the specific objectives of this study. Indeed, the guidelines of this strategic watershed management are based on critical analysis of the current situation and the definition of predictive interventions to revitalize natural ecosystems and to develop pastoral resources in order to support the local population needs of forage and fuelwood.

In this perspective, the spatial aggregate functions are used to identify priority areas by statistics combination of significant values in the GIS database obtained in previous work on this study linked to biophysical and hydrological environments. The hot spot analysis is used to calculate the Getis-Ord Gi* statistics for each feature related to erosion hazard zoning from neighboring entities in spatial data set [6].

3.1 Mapping of erosion susceptibility

The input data preparation and spatial analysis of projected actions to control soil loss hazard are performed in the GIS environment (ArcGIS 9.3). Thus, the biophysical and hydrometeorological data assessment is based on empirical models to produce decisional maps of the priority areas to be developed.

3.1.1 Biophysical and hydrometeorological data

The data used for the soils susceptibility analysis are divided into five groups of explanatory variables (R, K, LS, C, and P). These are climatic, geomorphology, topography (gradient and length slopes), geological and geomorphological data, hydrographic parameters (river density, distance to streams), and soil occupation. Thus, thematic maps are produced by geoprocessing of information obtained.

The map of climatic aggressiveness is extrapolated from climatic data available in the stations characterized by long observation periods more than 20 years. Therefore, the topographic parameters (LS) are derived from the Digital Terrain Model DTM Aster, and planimetric and altimetric accuracies are, respectively, 30 and 20 m. The interpretation of the soil characteristics is used also to classify soils in the Wischmeier Abacus and to approach erodibility factor [7, 8]. Furthermore, the land cover map is extracted from SPOT satellite images (resolution is 20 m) combined with recent Landsat ETM+ imagery through the supervised classification method and field observations.

3.1.2 Decisional maps

Soils susceptibility assessment corresponds to the spatial occurrence of soil loss (number of representative pixels) that has taken place under the impact of local environmental conditions. Thus, the analytical approach adopted is based on simulation models integrated with GIS tools in order to evaluate the behavior of the dependent variable (land loss location) from a spatial combination of predictive variables in homogeneous geomorphic units (pixels). The soils susceptibility is simulated by the Universal Soil Loss model [9, 10], considered the most robust approach for spatial assessment of the soil erosion hazard (A). Moreover, the basic hypothesis is that the potential soil loss will be triggered under the same conditions as in the past.

$$A = R \times K \times LS \times C \times P \quad (1)$$

where A is the mean annual soil loss (t ha^{-1} $year^{-1}$), R is the rainfall erosivity factor ($MJ.mm.ha^{-1}.h^{-1}.year^{-1}$), K is soil erodibility factor ($t.h.MJ^{-1}.mm^{-1}$), L and S are the slope-length and the slope-steepness factors (dimensionless), C is the cover and management factor (dimensionless), and P is the support practice factor (dimensionless).

Second, hydrometeorological study is used for flood sites. The flood hazard is one of the most destructive natural hazards of the environment that can cause severe social as well as economic losses. In majority situations, the modern methods of passive and active flood protection provide rational protection for people and property within watersheds. Consequently, the refinement of probabilistic and technical methods is totally justified.

The flood analysis is based on the dependence applied in the Francou-Rodier model [11] and on the distribution function characteristic of the extreme values. This approach is the most widely used in Morocco, as a regional empirical formula that has the advantage of making the flow value, with the defined exceedance probability, dependent on the basin area function.

The possibility to estimate the biggest possible flood that could appear during the extreme conditions is the significant element of the estimation of the potential hazard. Therefore, the Gradex model determines the flood flow characteristics and the regional parameter (k_t) in a gauged station located in the Oued Beht watershed [12]. Subsequently, the data obtained are extrapolated to the other sub-catchments by Francou-Rodier method, based on the regional coefficient (k_t) calculated in the gauged station called Ouljet Soltane [11].

Below, the two significant formulas, which enable the estimation of the form of the maximum flows envelopes, are described:

$$\frac{Q(t)}{Q_0} = \left(\frac{S}{S_0}\right)^{1-\frac{k_t}{10}} \quad (2)$$

where Q(t) is the maximum flow value in the ungauged sub-catchment (m^3/s), S is the area of ungauged sub-catchment (km^2), Q_0: 10^6 m^3/s, S_0: 10^8 km^2, and k_t is the parameter of Francou-Rodier, which is a regional parameter in the right of the gauged station (called Ouljet Soltane station).

Therefore, the first step consists in calculating the Francou-Rodier parameter (k_t), by using the flow Q_A for a determined return period, in the Ouljet Soltane gauged basin, whose area is S_A. Considering the data available on the gauged basin, the flow Q_A is calculated by the Gradex method.

$$k_t = 10 \times \left[1 - \frac{\ln\frac{Q_A}{10^6}}{\ln\frac{S_A}{10^8}}\right] \quad (3)$$

where Q_A is the flood flow in the gauged sub-catchment (m^3/s) and S_A is the area of the gauged sub-catchment of Ouljet Soltane (km^2).

The final map is the result of geoprocessing by spatial crossing of information linked to soils degradation by natural erosion and flood power to contribute to El Kansra dam siltation. As a result, the production of this qualitative map is used to provide a systematic vision to identify priority areas with homogeneous environmental characteristics and to study the alternatives of development upstream/downstream.

3.2 Socio-economic analysis

The methodological protocol used to characterize the socio-economic aspect is based on survey data [2, 3]. Thus, the analysis of mechanisms essentially linked to lifestyle needs and household income is used to better understand the socio-economic vulnerability in the watershed. The aim is to prepare a reference situation for the future socio-economic or environmental project.

The surveys are conducted in homogeneous areas using some direct conversations with groups surveyed (focus-group), with a freedom to structure the interview to better understand the population profile, their real needs, and to identify constraints that limit wealth production (natural, financial and commercial constraints, land structures, and incomes). Moreover, spatial distribution of the farms (units to investigate) is selected using a stratified sampling plan with 5% error and 95% confidence level. The randomness of the villages (sampling units) is made from the list available in the general agricultural census [13]. Based on the number of households in the watershed studied (19,987 farmers), the number of farmers to be interviewed is 378 in 50 villages.

Consequently, in order to reduce the heterogeneity linked to utilized agricultural land (UAL), which represents a discriminatory factor for management techniques and income sources, the stratification is performed according to farm size, and three classes are selected (UAL < 5 ha, 5 ha < UAL < 15 ha, and UAL > 15 ha).

3.3 Potential consequences

The potential consequences are evaluated by an analytical approach based on the identification of the exposed elements and the assessment of their vulnerabilities. In this approach, the potential damages are not expressed in numerical values but in hierarchical classes (qualitative assessment). The consequence typology differs: (1) direct structural damages (CS) affecting the land goods and the El Kansra dam and (2) direct functional damages (CF) related to disruption of agricultural activities with local and immediate consequences.

The consequence assessment is a fundamental part of erosion risk analysis. Thus, the various components of the vulnerability are structured according to a decreasing exponential function. Moreover, the vulnerability analysis is based on the observation protocol of damage, original and reproducible, applicable to the soil loss analysis due to past erosion events. The erosion cost is defined by the difference between the initial net revenue per hectare and the net revenue with the effect of erosion:

$$R_t = R_0 e^{-xp(t)} \tag{4}$$

where R_t is the yield in the year t (t/ha), R_0 is the initial yield (t/ha), x is the damage coefficient (yield loss parameter), and p(t) is the cumulative land loss in the year t (t/ha).

The vulnerability input is based on the results of socio-economic surveys describing the current yields (or revenue) and the latest census data available in the Office of the High Commissioner for Planning (HCP), the government agency in charge of producing statistics [3, 14–16]. Thus, the damage process typology helps to prioritize the consequences classes: low (C1), moderate (C2), high (C3), and very high (C4).

3.4 Erosion risk assessment

On the technical side, the terms "risk" and "hazard" are linked to each other but should be clearly distinguished. The risk mainly signifies a probability of the occurrence of (negative) impacts and expected losses resulting from a given hazard to a given element at danger or peril, over a specified time period.

Therefore, the purpose is to hierarchy the erosion menace that compromises land goods, human activities, and property of people. Thus, the analysis of the soil degradation levels obtained allows to prioritize the susceptibility classes: low (S1), moderate (S2), high (S3), and very high (S4).

In addition, the spatial combination of susceptibility (S1 < S2 < S3 < S4) and potential consequence (C1 < C2 < C3 < C4) are translated into risk classes using a correlation matrix of double entries [17]. Consequently, the erosion risk classes are prioritized in order to guide planning decisions (**Figure 3**): low (R1), moderate (R2), and strong (R3).

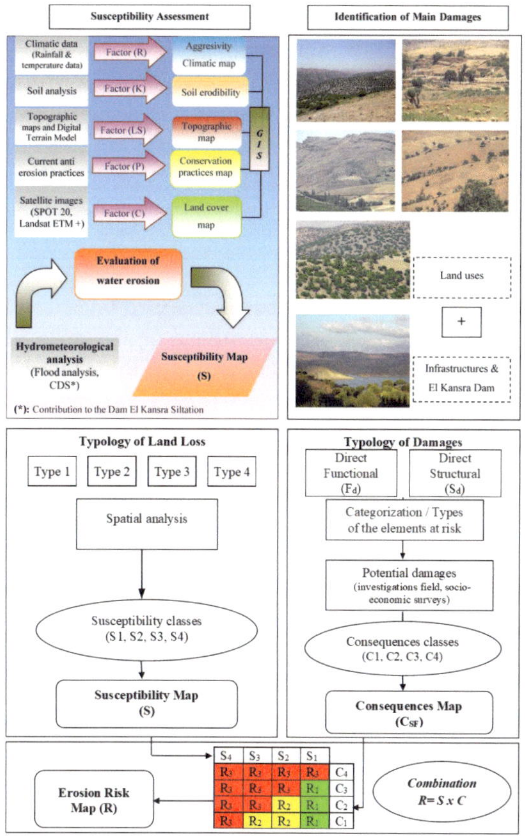

Figure 3.
The methodological flowchart to identify priority areas and evaluate soil erosion risk.

3.5 Watershed management plan

Using the analysis of the existing opportunities in Oued Beht watershed, the present master plan is based on an action program focused on erosion control. Indeed, the identification of the package of management actions is based on the diagnosis results of biophysical and socio-economic backgrounds. Thus, the management approach of priority areas is based on operational actions (biological and technical treatments) that are compatible with the intrinsic characteristics of the watershed studied. In this sense, the objectives that promote the action plan are as follows:

- Bioengineering techniques for soil erosion protection and slope stabilization to conserve natural resources upstream and to protect El Kansra dam against siltation.
- Reconstruction of degraded ecosystems to promote biodiversity conservation.

3.5.1 Strategic planning

The long-term planning is used to establish the framework and key elements of Oued Beht watershed and to reflect a clear vision created in an open process. Guidelines for the many departments which will draw up specific plans will be established. Thus, the key elements are reviewed for potential effects with uniform land uses (agricultural, rangelands, and forest).

The interventions program includes the conservation actions and environmental rehabilitation. The measures selected are grouped into the following categories: agricultural land use, rangelands management, forest management, river system protection, and ravines treatment.

3.5.2 Priority planning

The Hot Spot Analysis tool calculates the Getis-Ord Gi* statistic for each feature in a spatial data. The resultant "z-score" (standard deviation) tells us where features with either high or low values cluster spatially. This tool works by looking at each feature within the context of neighboring features. A feature with a high value is interesting, but may not be a statistically significant hot spot. Thus, to be a statistically significant hot spot, a feature will have a high value and be surrounded by other features with high values as well. The local sum for a feature and its neighbors is compared proportionally to the sum of all features; when the local sum is much different than the expected local sum, and that difference is too large to be the result of random chance, a statistically significant "z-score" results.

The Gi* statistic returned for each feature in the dataset is a "z-score." For statistically significant positive "z-score," the larger the "z-score" is, the more intense the clustering of high values (hot spot); and for statistically significant negative "z-score," the smaller the "z-score" is, the more intense the clustering of low values (cold spot).

In this study, the use of the statistical method "Getis-Ord Gi*" allows the analysis of each entity (pixel) in relation with neighborhood in the spatial dataset [6]. The nearest neighbor analysis is based on comparing the distribution of the distances from each data point to its nearest neighbor in a given dataset with a randomly distributed dataset.

Indeed, this statistical approach tells us if we may reject or not the null hypothesis CSR (complete spatial randomness) that expresses the absence of spatial correlation between the following events: 1) significant soil loss and degraded vegetation cover and (2) soil erosion and steep slopes. Thus, the results, expressed in "z-score" (standard deviation) and "p-value" (independence probability), are used to measure the statistical significance of spatial autocorrelation (**Figure 4**).

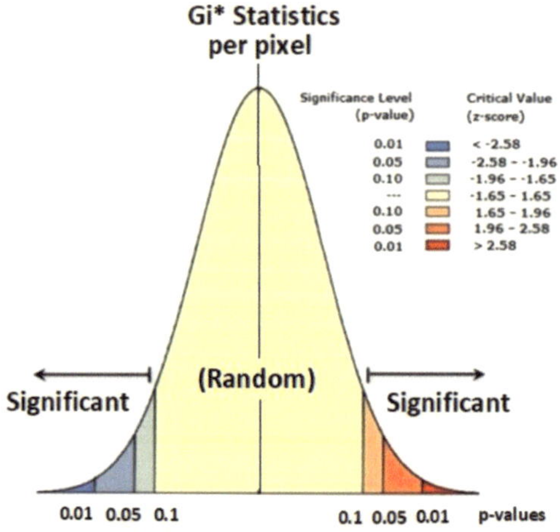

Figure 4.
Distribution of spatial autocorrelation indicators (adapted from [18]).

Furthermore, for confidence level 90%, if the z-score obtained is between −1.65 and +1.65, the probability of independence (p-value) will be automatically higher than 0.10 and the null hypothesis of independence is not rejected [18]. Thus, the biological actions are programmed in areas with strong spatial autocorrelation (hot spot) between erosion hazard and vegetation cover; and technical measures correspond specially to areas with high spatial autocorrelation between the natural hazard and topographic factors that will be modified by the action plan.

4. Results and discussion

4.1 Biophysical factors analysis

4.1.1 Topographic context

The topography of the Oued Beht watershed is the result of factors with a combination involving topographic effect of altitudinal amplitudes, exposure, slope gradient, and slope length.

4.1.1.1 Hypsometric analysis

The spatial analysis of the digital terrain model (DTM) shows that the watershed has a regularly altitudinal distribution along its elongated form. Thus, altitudes classes obtained follow a decreasing gradient, from upstream to downstream, in perpendicular bands to the axis, which coincides with the flow direction of Oued Beht (**Figure 5**).

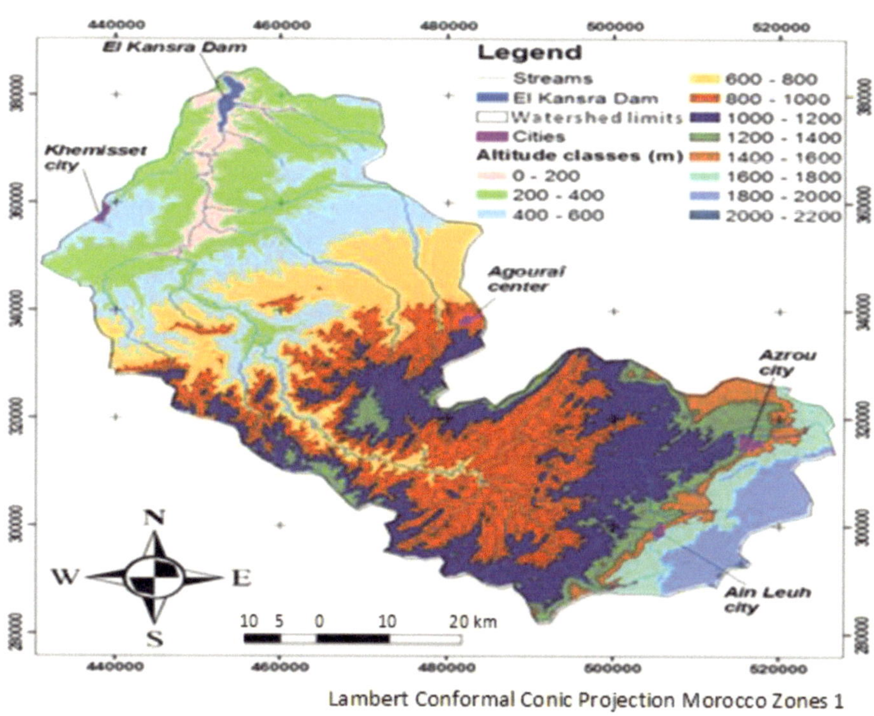

Figure 5.
Hypsometric map.

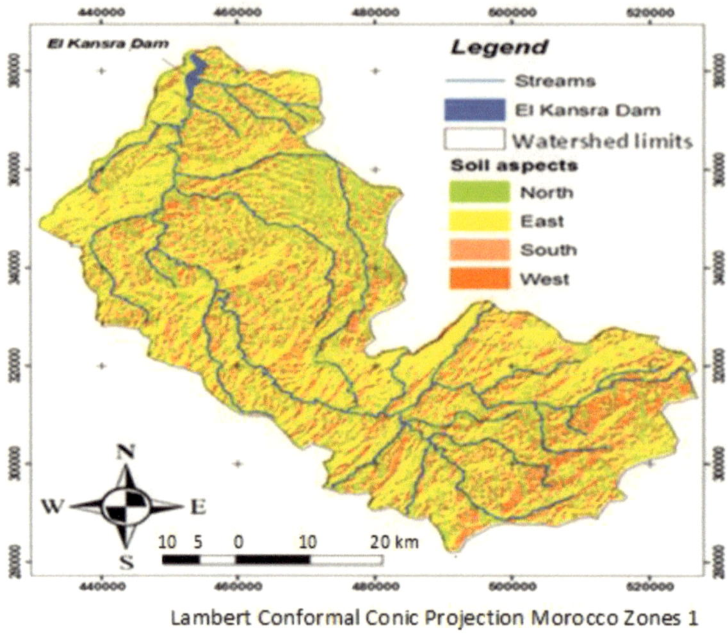

Figure 6.
Aspect map.

The watershed presents a high altitudinal range, between the highest point 2121 m and the lowest point 108 m, which coincides with the level of the El Kansra dam. Thus, the total length midline crossing the watershed is 177 km, and the altitude difference of 2013 m represents a real hydrologic indicator that promotes erosive process.

4.1.1.2 Soil aspects

The aspect map is used to establish the slope exposure of the watershed and to give an idea about the relief forms and the cover land. The distribution of soil aspects shows that east facing slopes dominate, particularly at upstream part of Oued Beht watershed (38%). However, the other exposures are equal, almost 20%, while specifying that the investigations show that the north and west slopes present a humid character. Furthermore, the areas representing a flat field (with 0% of slope) are limited and localized mainly in the small depressions or hilltops (**Figure 6**).

4.1.1.3 Slope gradient analysis

The DTM spatial analysis shows that the low slopes (less than 15%) are dispersed and occupy more than half of the watershed (57%). Thus, steep slopes are concentrated in central and upstream areas. The map of the slope length classes gives an indication of the transport distance traveled by soil particles detached. The slope lengths distribution shows that almost half of the watershed (55%) is less than 1000 m with a majority (30%) lower than 500 m (**Figure 7**).

In this sense, the digital terrain model (DTM) is the main source of data for the extraction of many parameters used such as slope lengths, direction of flow of

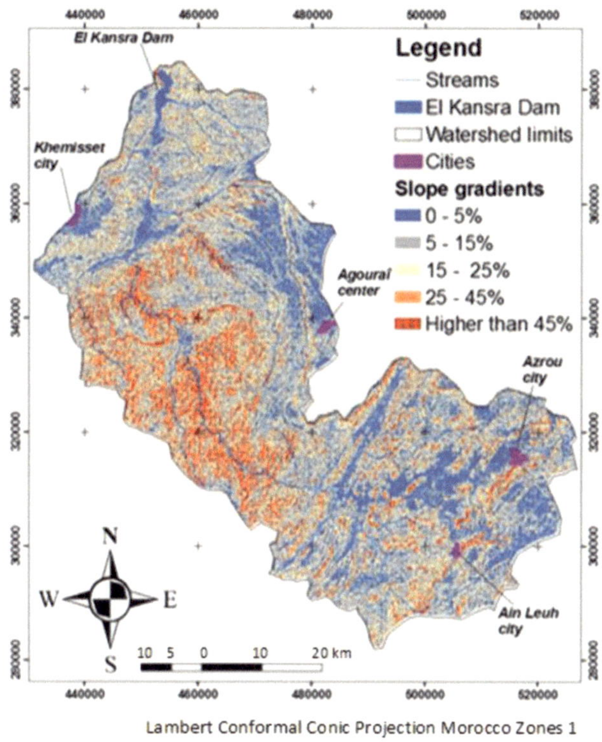

Figure 7.
Slope gradient map of the study area.

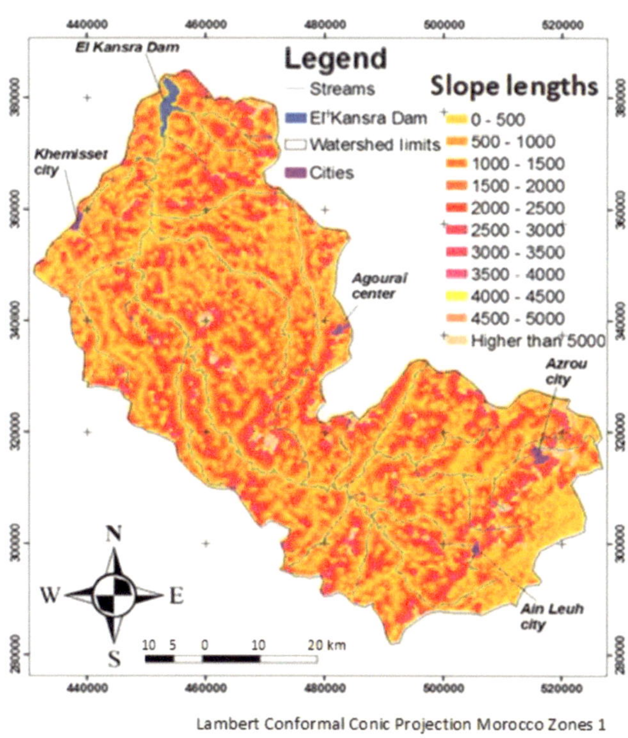

Figure 8.
Slope length map.

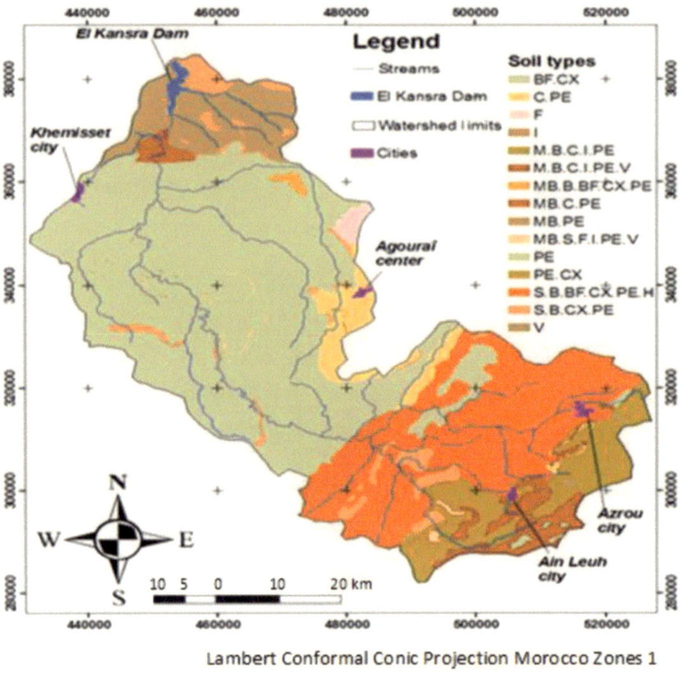

Figure 9.
Soils map.

water, topographic index, etc. The spatial distribution of the slope length classes is heterogeneous, and no zone is characterized by a single slope length class (**Figure 8**). Also, their importance decreases to a minimum corresponding to the class exceeding 5000 m with only 1%.

In conclusion, the topographic factor analysis reveals the combination of slope length effects with slope gradient characterizing Oued Beht watershed.

4.1.2 Soil resources

The soil analysis in Oued Beht watershed shows a strong dominance (45%) of slightly developed soils (PE). This soil type is dispersed and used not only for agriculture and forestry but also in rangelands. Moreover, Brown soils (B) and Forest Brown (BF) soils are concentrated at the upstream where the forests are developed (6%). Thus, this kind of soil is enriched by the litter decomposition (**Figure 9**). Specially, the poor soils, characterized by the bedrock outcrop, are located near El Kansra dam and at the extreme south of watershed (upstream).

In conclusion, the watershed soils analysis shows the diversity and heterogeneity of pedogenesis factors. Thus, this diversification of soils is mainly due to bedrock types and their degree of friability, morphology, topography, climate aggression, and land use (**Figure 10**).

4.1.3 Hydrometeorological analysis

The geographical distribution of climate stations selected presents good spatial coverage and long periods of observation that allow an eminent climate analysis in Oued Beht watershed. The weather stations used to characterize the thermal regime and deduct bioclimatic classes are the stations of El Kansra, Khemisset, and Ifrane (**Figure 11**). In this way, the continentality is quite significant with a neat decrease

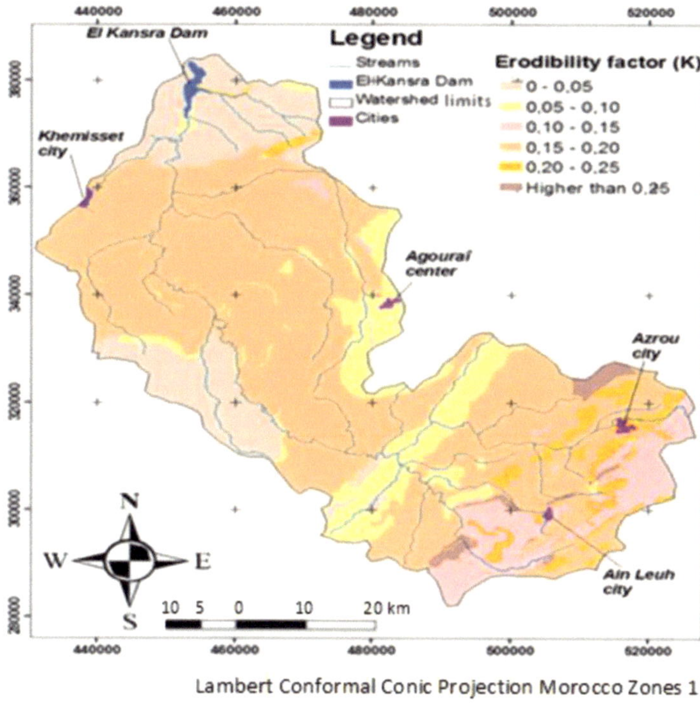

Figure 10.
Soil erodibility distribution.

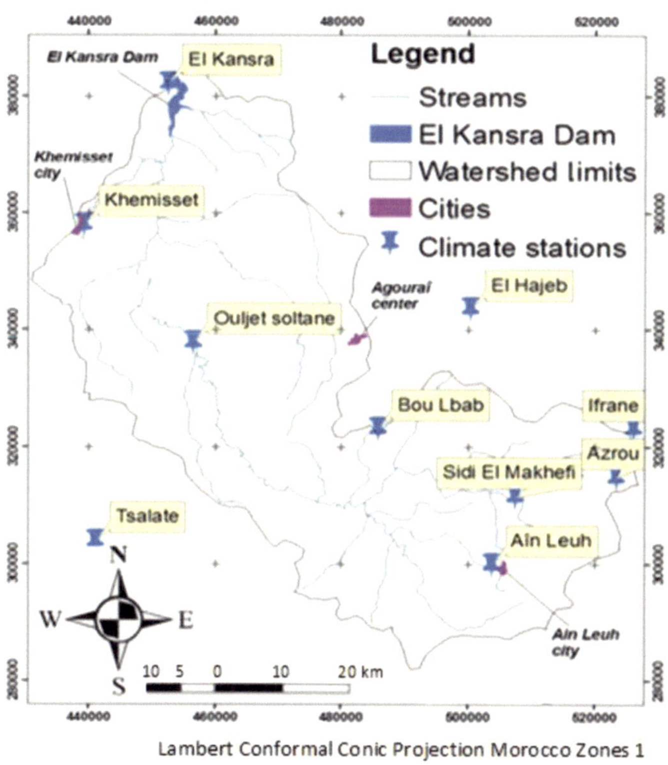

Figure 11.
Distribution of climate stations.

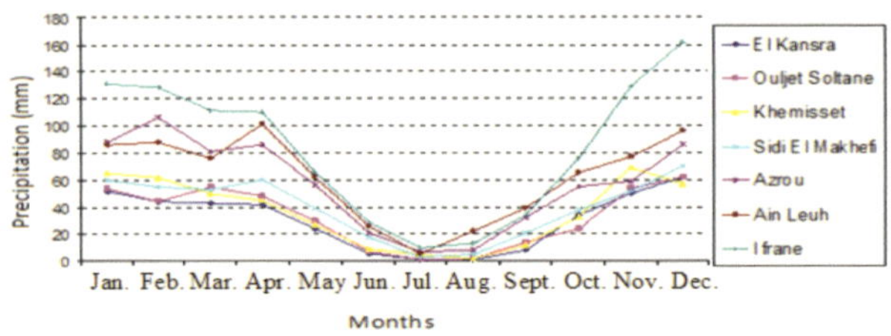

Figure 12.
Distribution of average monthly rainfall data by station.

in temperature associated with increasing altitude. Moreover, the thermal regime is characterized by average temperatures that vary between 10°C in the east and 26°C in the north and north-west. Thus, the studied watershed is influenced by altitude and latitude factors.

On the other hand, the rainfall regime is irregular and the rainy period is concentrated between October and May (**Figure 12**). As a result, the precipitation distribution analysis shows that the watershed has a rainy winter and a dry summer period. Therefore, the upland areas (mountains) are wetter than the areas that are close to the sea. Thus, the altitude effect on rainfall (R-factor) is more dominant than the approximation of the sea.

4.1.3.1 Bioclimatic synthesis

Bioclimatic data analysis is based on quotient Emberger index (Q2). This quotient, especially adapted to the Mediterranean regions, is based on the annual rainfall, the average maximum temperatures of the warmest month (M °C), and the average minimum temperatures of the coldest month (m °C) [19]. Thus, Oued Beht watershed is characterized by several bioclimatic architectures:

- In the north and northwest, the climate is semi-arid with temperate winter.
- In the center, the climate is sub-humid with temperate winter.
- The east of the watershed presents a humid climate with cold winter.

In addition to the data mentioned above, linked to altitudinal impact (2013 m), the watershed hydrological behavior is conditioned also by the bioclimatic changes affecting inevitably the nature of the developed vegetation, the resilience of different ecosystems, and intensity of erosion hazard.

4.1.3.2 Rainfall aggressiveness (R)

The rainfall erosive power, or the rainfall erosivity factor (MJ mm ha^{-1} h^{-1} $year^{-1}$), is calculated by the application of the formula using data of average monthly and annual rainfall in the selected meteorological stations [17, 20]. Thus, the rainfall aggressiveness values (R) are between 64 and 130, respectively, recorded at El Kansra and Ifrane stations. Moreover, in the east, the rains are more aggressive than in the north and west. Also, the upstream area shows the higher rainfall aggressiveness indexes (**Figure 13**).

In conclusion, the rainfall aggressiveness, associated with the heterogeneity of the rainfall distribution, is spatially variable and adheres to erosion processes [21, 22].

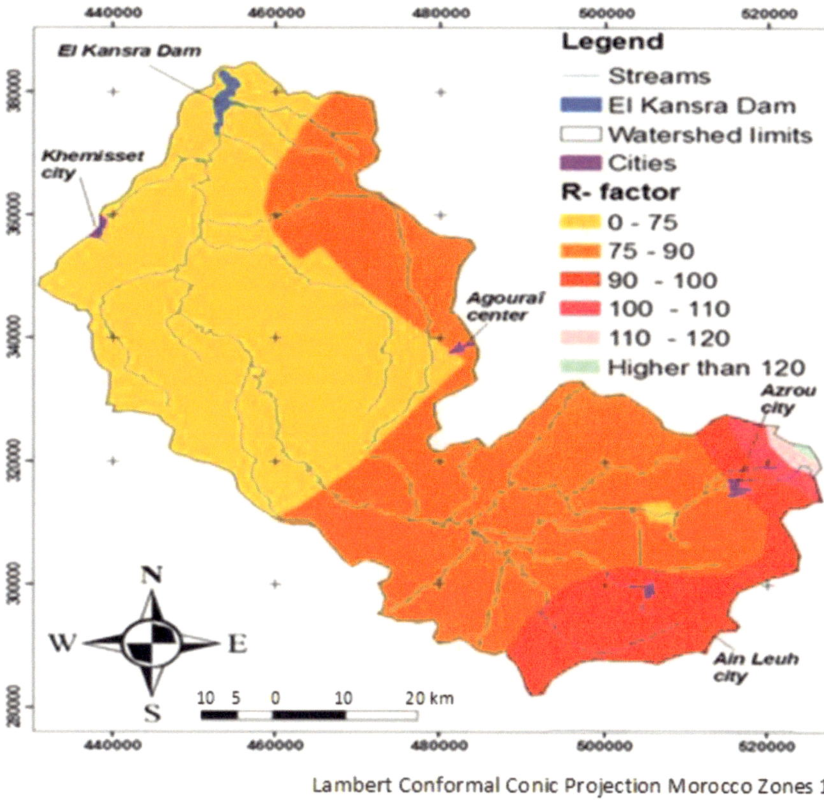

Figure 13.
R-factor distribution.

4.1.4 Land uses

The watershed has a variety of land uses related to the bioclimatic variation and topo-edaphical diversity. Thus, the rangelands area is the most common type of land cover (44%). Forests represent second place with 29%, reflecting the pastoral character of the watershed (**Figure 14**). Moreover, the forestry formations are concentrated mainly in the central and upstream. Furthermore, we note the presence of unplanted lands, covered by rocks, which are generally concentrated near the dam El Kansra.

4.1.5 Vegetation index

Normalized difference vegetation index (NDVI) is the most common measurement used for measuring vegetation cover. NDVI calculation allows to quantify vegetation by measuring the difference between near-infrared "NIR" (which vegetation strongly reflects) and red light (which vegetation absorbs), according to the following formula:

$$NDVI = \frac{(NIR - Red)}{(NIR + Red)} \tag{5}$$

The NDVI will be computed temporally to understand the change of land cover during the study period. It ranges from values −1 to +1. Thus, very low values of NDVI (−0.1 and below) correspond to barren areas of rock, sand, or urban/built-up. Zero indicates the water cover. Moderate values represent low density of vegetation (0.1–0.3), while high values indicate vegetation (0.6–0.8).

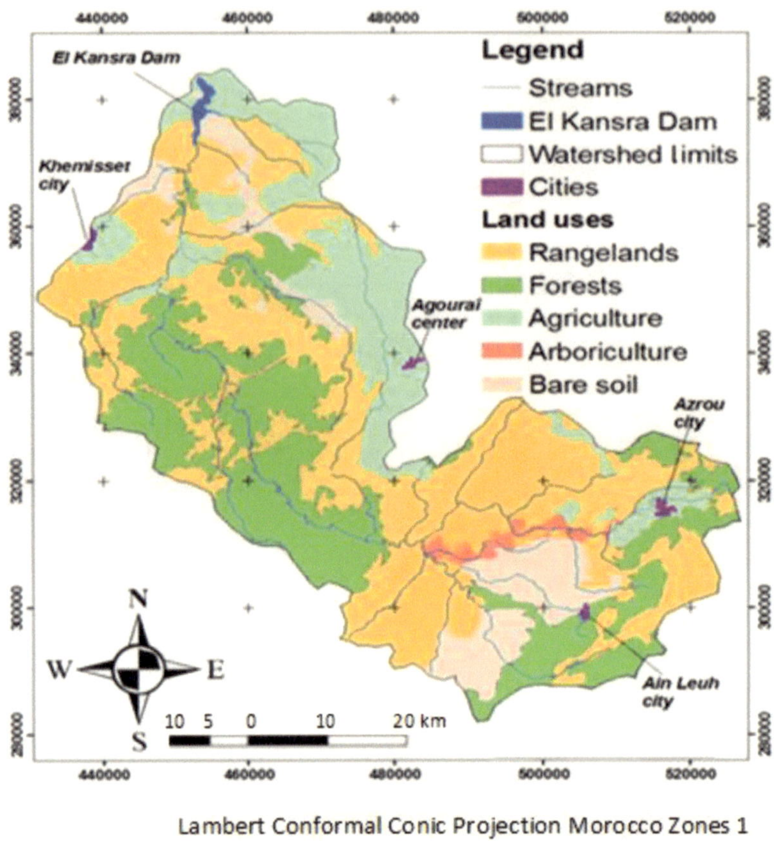

Figure 14.
Land use map.

The results obtained from the NDVI analysis show that the recovery rate is characterized by dominance of the low class, grouping generally rangelands and crop fields. Thus, both classes "low" and "very low" represent 72% of the total area. Consequently, this indicator reflects the low coverage capacity even if the land cover is almost complete and denuded soils rate is only 9.5% (**Figure 15**).

Moreover, agricultural lands are specially based on cereals and annual crops with short growing cycles. Thus, the rangelands consist of perennial grass vegetation with short development cycle. In conclusion, the land use duration is short, especially during periods of heavy rain.

The analysis of vegetation cover (C-factor) also confirms the low recovery rate. Thus, more than half of the watershed (55%) has a C-factor exceeding 0.5 and 72% has values greater than 0.2 (**Figure 16**). Therefore, these results are consistent with the biophysical analysis describing the low recovery rate.

In conclusion, this factor has a detrimental effect on the erosion process by promoting the sediments production in low soil coverage, and especially if it is combined with other determinant factors.

4.1.6 Hydrological behavior

The establishment of the hydrological system map, based on DTM spatial analysis, allows to determine the rivers' directions and the accumulation of their flow (flow accumulation). Indeed, the river system obtained is ramified along the entire watershed. Thus, it consists of the main stream named Oued Beht, which

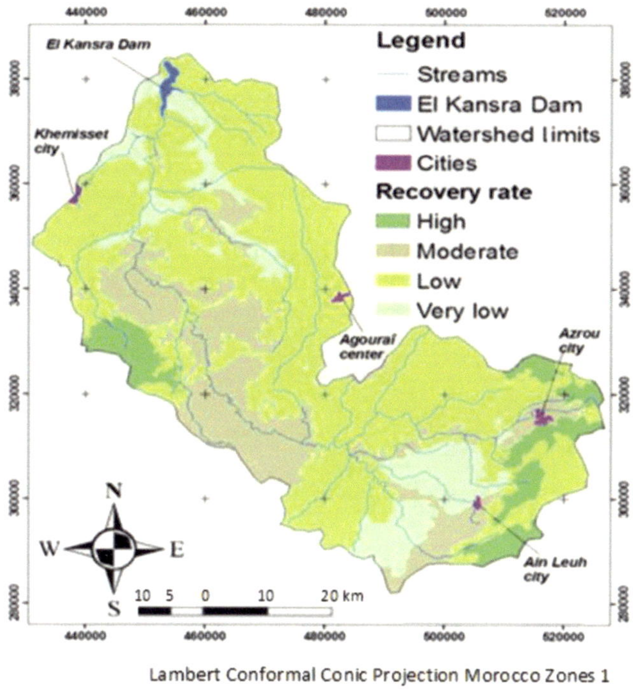

Figure 15.
Normalized difference vegetation index map.

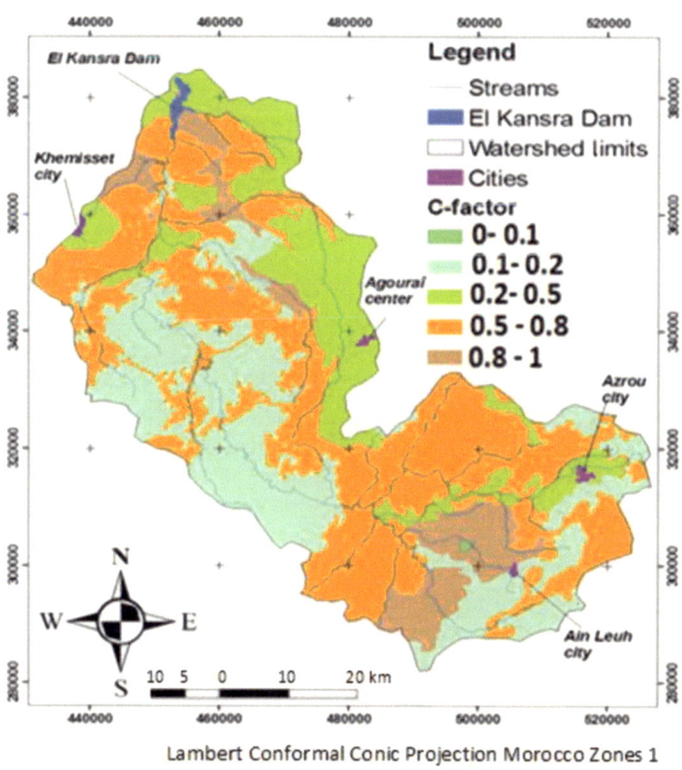

Figure 16.
C-factor distribution.

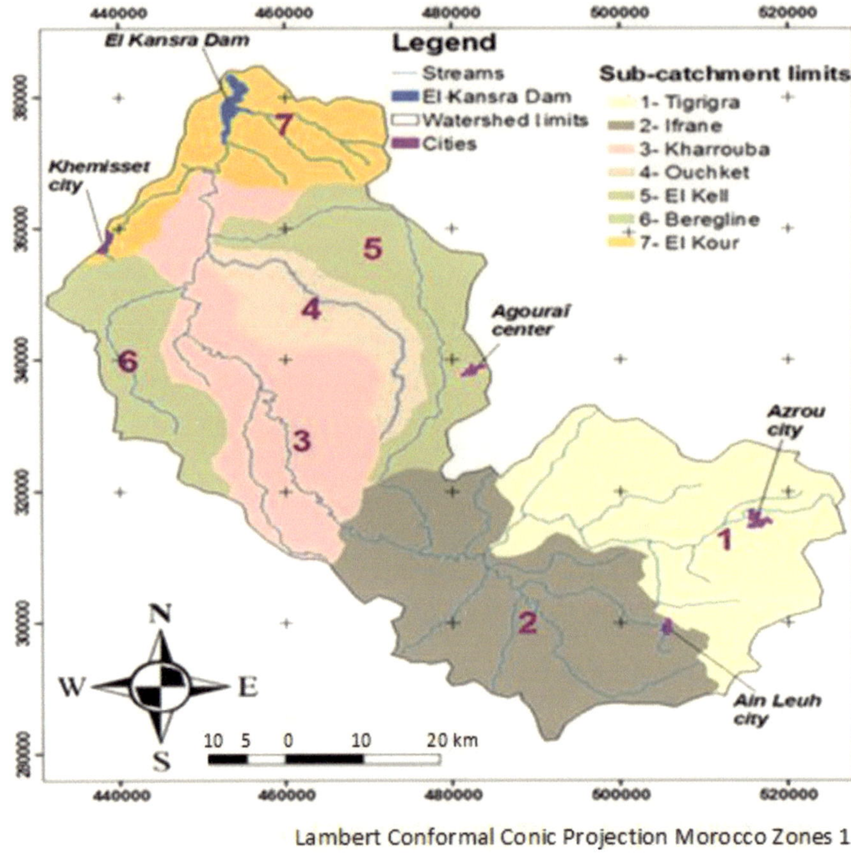

Figure 17.
Sub-catchment delimitation.

is powered by the waters of several tributaries: Beht, Tigrigra, Ifrane, El Kell, Ouchket, Kharrouba, Beregline, and El Kour (**Figure 17**).

4.1.6.1 Drainage density

The surface drainage in the Oued Beht watershed is assured by a hierarchical arsenal of rivers. Thus, the density is influenced by its topo-geological structure and relief. Indeed, the river system is characterized by the importance of its elements, since their original ramifications upstream, domiciled in the Middle Atlas chain, to the main collector which is the El Kansra dam. Therefore, the river system is characterized by a total length about 667.77 km and an average density 0.16 km/km^2.

4.1.6.2 Concentration time

The time (t_c) that is necessary for the farthest water particle to arrive at watershed outlet, is estimated by the formula of Passini [23], which is presented as follows:

$$t_c = 0.108 \times \frac{(S \times L)^{\frac{1}{3}}}{I^{\frac{1}{2}}} \tag{6}$$

where t_c is the concentration time (h), S is the surface of the sub-catchment (km^2), I is the average slope of the sub-catchment (%), and L is the length of the longest path flow (km).

Therefore, the concentration time (t_c) is relatively low at the majority of sub-catchments (SBC) and varies from 2:30 hours (in SBC/Kharrouba) to almost 5 hours (in SBC/Tigrigra).

In conclusion, the elongated form of the Oued Beht watershed and the low concentration time for the majority of sub-basins are favorable conditions for the development of flood and river flows that cause sediment deposits in the stream beds and El Kansra dam.

4.1.6.3 El Kansra dam siltation

The bathymetric data analysis implemented since the construction of the El Kansra dam is used to assess the quantity of soil loss which compromises the storage capacity and quality of water flow. Thus, for a period of 23 years (1981–2004), the average El Kansra dam siltation is 3 million m^3/year.

4.1.6.4 Floods study

The central objective is the prioritization of sub-catchments presenting high flood risk and soil erosion. The data linked to maximum flood flows are obtained by calculating the extreme gradient values (Gradex method) from the decennial flow in the reference station of Ouljet Soltane (**Table 1**). Thus, the hydrological analysis involved determination of design floods for a large number of sub-catchments by the Gradex method:

- First, samples of annual maximum daily rainfall were used to calculate parameters P0 and G of Gumbel distribution for the various raingauge stations (P0 = ordinate of origin and G = slope or gradex).

- The Gradex method was next used, with a daily time-step, applied to all stream gauging stations available. Thus, the pivot point was taken as $T^* = 10$ years. Conversion from daily discharge $Q_j(T)$ into peak discharge $Q_p(T)$ was done by considering the mean Q_p/Q_j ratio from a small sample of hydrographs.

- Lastly, the results obtained (**Table 2**) are synthesized using the following equation for calculating the maximum instantaneous flow $Q_p(T)$, for the return period T:

$$Q_p\,(T) = Q(T^* = 10 \text{ years}) + C_p * G_d * (u(T) - u(T^*)) \tag{7}$$

where G_d is Gradex flow, defined by the following formula:

$$G_d = G_p * S/(3.6 * t_c) \tag{8}$$

Return period T (years)	10	20	50	100	1000
Q (m3/s)	488	586	712	807	1121

Table 1.
Statistical adjustment of annual maximum flows, in the gauging station of Ouljet Soltane.

Streams	Area (km2)	Q p	Q p	Q p	Q p	Q p
		T = 10 years	T = 20 years	T = 50 years	T = 100 years	T = 1000 years
Tigrigra	909.37	241	294	364	417	597
Ifrane	1019.6	261	318	394	451	643
El Kell	487.2	154	190	238	275	401
Ouchket	326.53	115	143	181	210	310
El Kour	413.2	137	169	213	246	361
Kharrouba	798.12	219	268	333	382	549
Beregline	353.26	122	152	191	221	326

Table 2.
Flood flows Q_p of the principal rivers (m^3/s).

where G_p is the Gradex rainfall, S is the area of the watershed (km^2), t_c is the concentration time (h) (Eq. (5)), C_p is the pivot point, and u(T) is the variable of Gumbel.

For the other neighboring ungauged sub-watersheds, the application of Francou-Rodier formula gives the following results (Eq. (2)):

4.1.7 Potential erosion

Compared to Eq. (1), the potential erosion allows to evaluate the power of soils to produce sediments under the effect of rainfall and topological factors, without considering land cover (C-factor) and erosion control practices (P-factor). The crossing of thematic layers of rainfall aggressiveness (R), soil erodibility (K), and the topographic data (LS) is used to synthesize potential impacts according to the formula defined as follows [9, 10]:

$$E_p = R \times K \times LS \tag{9}$$

where E_p is the potential average annual soil loss (t/ha/year), R is the rainfall aggressiveness index ($MJ.mm.ha^{-1}.h^{-1}.year^{-1}$), K is the soil erodibility ($t.h.MJ^{-1}.mm^{-1}$), and LS is the topographical factor (dimensionless).

The results analysis shows that the potential average annual soil loss is 54 t/ha, and the average annual quantity is 23.25 million t/year. Moreover, the importance of soil loss differences between extreme values obtained (pixels) shows the power of eminent soil units to produce sediments under the rainfall aggressiveness [24].

Two-thirds of the Oued Beht watershed are characterized by soil loss quantity, which is less than 50 t/ha/year, and almost 30% corresponds to the potential erosion class between 50 and 300 t/ha/year. Thus, on the broken reliefs located in the upstream part (in Tigrigra and Ifrane sub-catchments), with steep slopes, generally exceeding 25%, the potential erosion is high with values that may exceed 200 t/ha/year (**Figure 18**). Moreover, these erodible areas are characterized particularly by high and medium soil friability.

Second, some areas near El Kansra dam present high values of the potential erosion exceeding 200 t/ha/year. These vulnerable sectors correspond mainly to northern sub-catchments with low altitudes (less than 400 m) with high soil friability. Therefore, the great erosive power of adjacent areas to El Kansra dam is a real danger involving the dam siltation and compromising its service life.

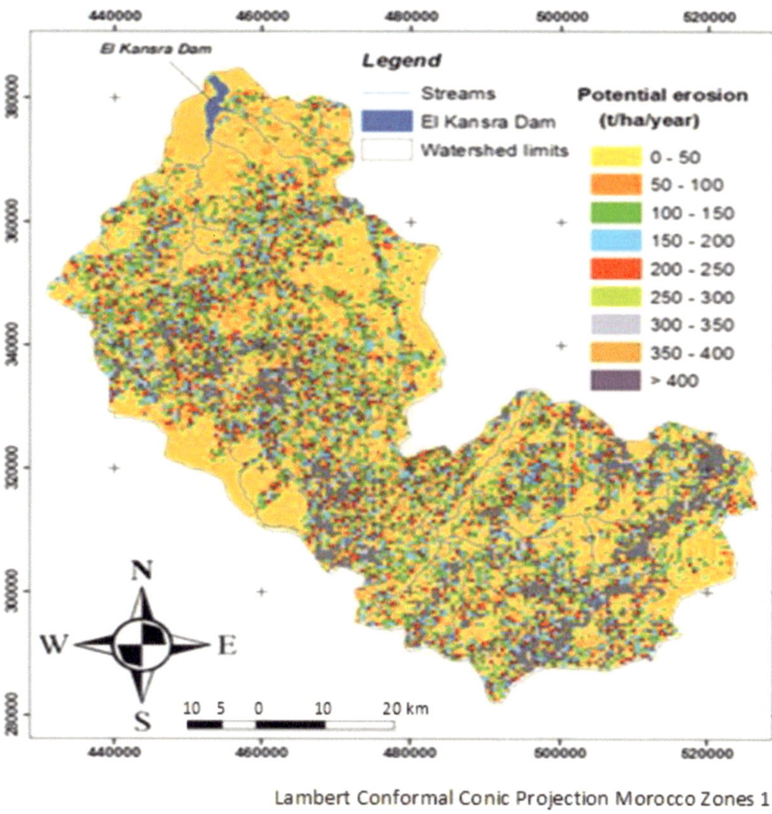

Figure 18.
Potential erosion map.

Furthermore, the investigative visits show that the upstream part is very sensitive to the potential erosion, but it should be noted, by location, the presence of medium and high levels of vegetation cover that can reduce the erosive potential.

The priority areas delimitation is performed through the spatial crossing of the specific degradation map, the map of sub-catchment contribution to dam siltation, and flood generation. This analysis is further developed by the socio-economic vulnerability map (**Figure 18**). Thus, we note that the results obtained reveal that the majority of areas identified and delineated as priority areas are occupied generally by soils with strong erosion risks. Consequently, the vulnerability linked to soil degradation characterizes 32% of Oued Beht watershed (**Figure 19**).

In conclusion, 24 rural communes know high contribution to dam siltation and include areas with high erosion risks and high poverty level. Therefore, urgent biological and technical actions are needed in this region to control erosion impact [25]. Therefore, these rural communes are concerned by action plans linked to land uses (agriculture, livestock, and forests).

4.2 Mapping erosion susceptibility

The hazard zoning obtained and the analysis of cumulative curves (number of pixels) define four susceptibility classes in the Oued Beht watershed:

- Low susceptibility (S1): The start of the erosion is negligible in almost half of the watershed (44.5%). In fact, local conditions contribute to the stability of

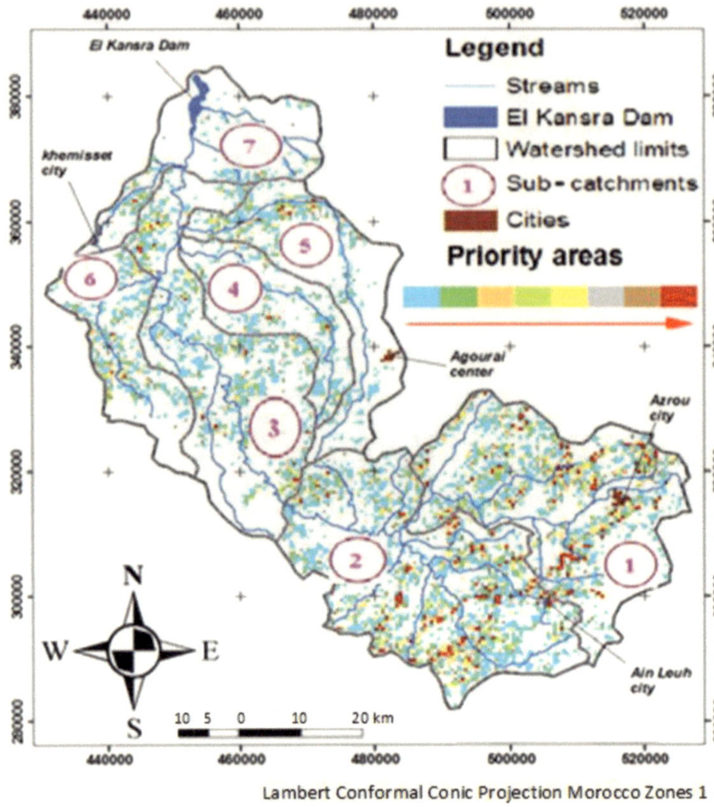

Figure 19.
Distribution of vulnerable areas.

the land. Gradients of the slopes are very low (lower than 5%) on agricultural land which is well maintained and well drained.

- Moderate susceptibility (S2): Local environmental conditions are also favorable to the onset of low land loss in almost a quarter of the watershed (24.4%). It is protected by forest areas and the slope gradients are low to moderate (5–25%). However, the abandonment of the reservation land or the local presence of slope failure could lead to destabilization.

- High susceptibility (S3): Local environmental conditions are favorable for triggering erosion (11.4%). It is rangeland and unprotected forest formations located on moderately degraded soils and characterized by poor soil drainage techniques. The degrees of slopes are moderate to strong (25–45%).

- Very high susceptibility (S4): The possibilities of the start of erosion are strong and the local environmental conditions are very favorable for that in 19.7% of the watershed. Soils are severely degraded, poorly maintained, and managed. The general appearance is marked by the absence of vegetation or forests. Thus, the erosion is very active with a significant soil loss with strong slope gradient (more than 45%).

In conclusion, this exploratory procedure shows amply the system capacity to generate automatically the hazard zoning. Almost a third (31%) of the Oued Beht watershed presents high to very strong susceptibility. The four hazard levels can

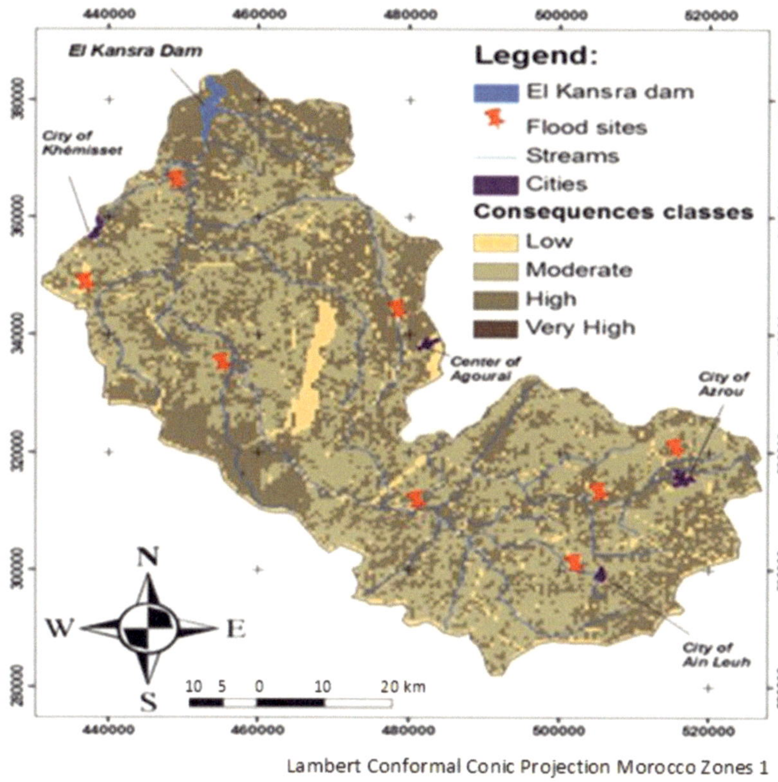

Figure 20.
Potential consequences map.

be combined with vulnerability with four levels. This integrated analysis would produce risk maps, or rather the existing deficit protection.

4.3 Potential consequences analysis

The analysis of the socio-economic vulnerability of the watershed is based on the assessment of damage related to the effect of past erosive events on the profitability of soil resources and the income of the farmers surveyed in this study. Thus, the preparation of input data is based on the results of socio-economic surveys describing the decline in land yield year after year (income loss).

As a result, the yield loss parameters that tell us the annual cost of erosion are defined by the differences between the net initial income per hectare and the net income with the effect of erosion (Eq. (4)).

In addition, the results of the socio-economic surveys show that the local economy is mainly represented by the primary sector (farming and poly-culture). The structural and functional damage map (CSF) describes the combination of damages due to land loss and El Kansra dam siltation that affect human activities. Therefore, the potential damage map (**Figure 20**) is obtained from the qualitative assessment of the state of land degradation (the importance of sheet, rill, and gully erosion) and this, to structure the cost of erosion and to highlight the homogeneous areas of vulnerability. Indeed, the analysis of cumulative curves (number of pixels) has identified four consequences classes for the Oued Beht watershed.

- Low consequences (C1): Minor damages to these lands are obsolete (1%) and hazard causes as much damage to human activities.

- Moderate consequences (C2): Mild to serious damage to soils and to infrastructures, which are characterized by half of the watershed (49%), mainly in the south watershed (upstream side) and partly downstream. Moreover, disruption of socio-economic activities is also moderate.

- High consequences (C3): Moderate to severe disturbances of human activities. Thus, strong and direct consequences are confined in space, but can be felt over the agricultural seasons; also, they represent almost half of Oued Beht watershed (50%). These consequences are partly located in the north watershed (downstream) and mainly around the El Kansra dam but locally extending toward the center and south.

- Very high consequences (C4): The very strong damage is minimal and negligible (0.02%); this kind of erosion events would exceed the human capacity and prevention authorities concerned.

4.4 Erosion risk management

The risk map (**Figure 21**), derived from a spatial combination of susceptibility and potential consequences classes, shows that high-risk areas (R3) are developed on 6% of the territory. These sites identify the major risks and disruptions of human activities. The warning areas correspond to areas with high consequences (C3), located immediately in the upstream side and locally to the center, presenting a very high to moderate

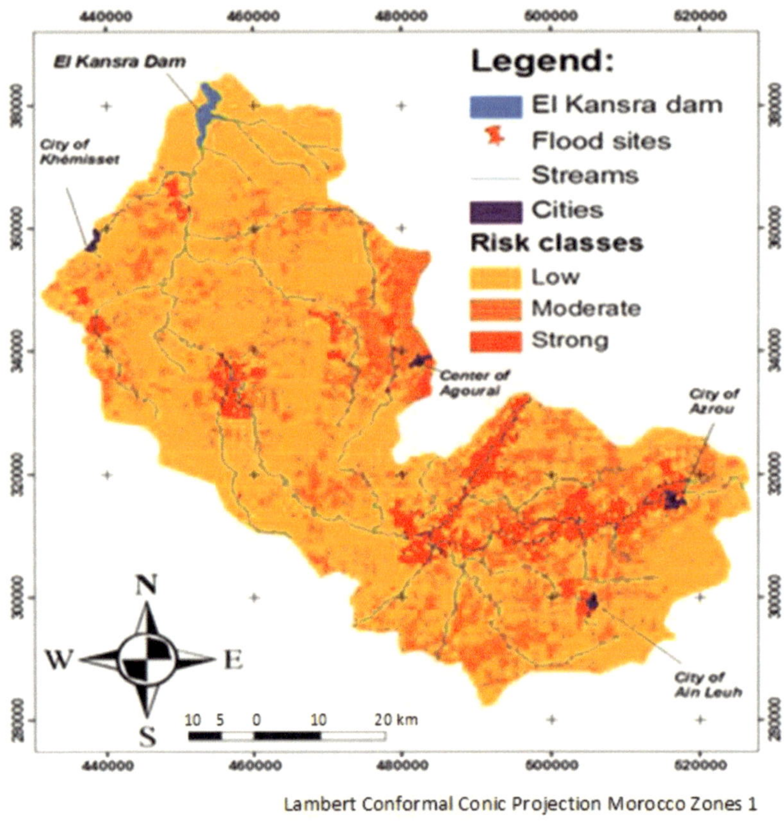

Figure 21.
Natural erosion risks.

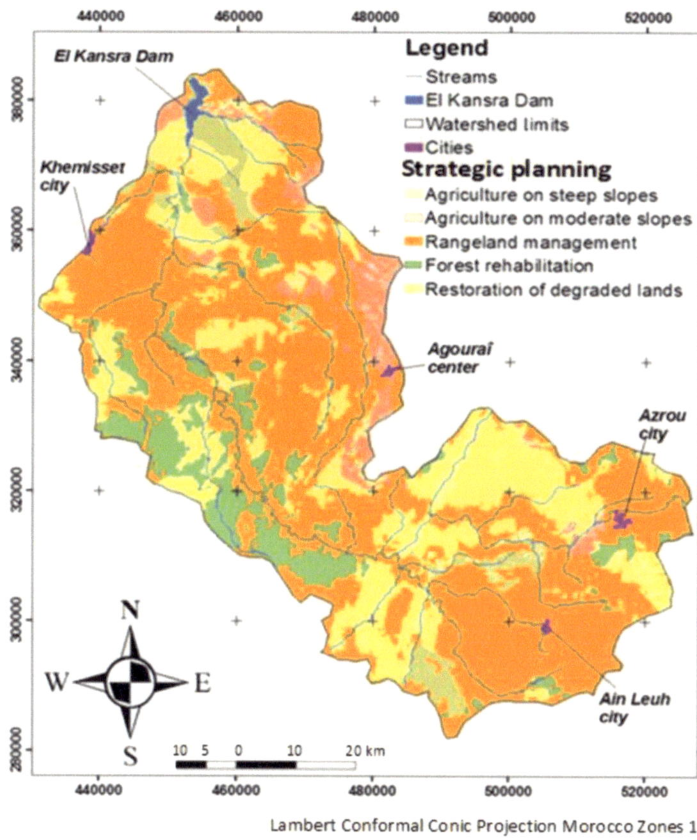

Figure 22.
Master plan of the Oued Beht watershed management.

susceptibility (S2, S3, and S4). Thus, appropriate precaution measures must be established (protected areas) and a risk prevention plan (RPP) must be implemented.

Elsewhere, outside large spaces present a low risk (R1) on 72% of watershed, representing the concept of acceptable risk. The risk level is moderate (R2) in 22% of the watershed (e.g., steep slopes but with low to moderate consequences). This menace presents a moderate disruption to human activities and serious damage to infrastructure including El Kansra dam. In conclusion, if improper resource management is implemented, this part of the watershed affected by moderate risk (22%) can be aggravated. Therefore, the potential risk can meet 28% of the watershed. Certainly, the development of management scenarios can complete this mapping study to improve the prevention of erosion risk.

4.5 Master plan for strategic planning

The formulation of Strategic Action Program (SAP) is based on the results of erosion risk mapping (**Figure 21**) to identify priority areas, where measures against soil erosion or reservoir siltation should be taken. The approach used is translated into operational actions (biological and technical), which are compatible with the intrinsic possibilities of the studied watershed (**Figure 22**). Thus, the Strategic Action Plan aims to achieve the priority objectives as follows:

- Measuring and monitoring soil erosion in order to preserve the natural resources and control El Kansra dam siltation.

- Flood analysis in order to reduce the flood risks with implementation of technical actions and Hydro-Agricultural Infrastructure Protection Plan (HIPP) including the El Kansra dam and the land goods.

- Implementation of biological actions, which consist of plantation and reforestation in degraded areas.

4.5.1 Agricultural land management

The agricultural development board assists rural households in the Oued Beht watershed to develop their agricultural business according to the lithological formations, and topographical and climatic constraints. Indeed, the agricultural lands, including arboriculture, cover an area of 74,577 ha, nearly 17% of the watershed area. Moreover, the operating systems are basically extensive with the cultivation of a maximum surface whatever the slope (even in the steep slopes).

On the other hand, the production systems adopted are characterized generally by inappropriate farming practices that promote soil erosion. Thus, the socio-economic study shows that 96% of rural population is conscious of the water and soils degradation.

In this sense, the selected actions aim to achieve a progressive evolution of production systems and land uses in accordance with soils vocation, with limitation of annual crops on steep slopes, the development of arboriculture, and improvement of forage production for livestock. Therefore, the implementation of actions mentioned below will lead to the increase of agricultural incomes and the establishment of a space management model to ensure local sustainability according to the following practices:

- Low to medium slopes (0–15%): The biophysical data analysis shows that the lands with low to medium slopes (0–15%) are subject to an erosive process generally manifested by sheet, rill, and rarely gully erosion. Thus, the aggressive rainfall and inappropriate farming practices (soil tillage in the direction of the slope, overgrazing) are the main factors that increase soil erosion.

The correctional measures include improving productivity through appropriate use of culture techniques. Thus, on low to medium slopes, the soil tillage must follow the contours and be combined with cultures in alternate bands.

In conclusion, to maintain this type of soil vegetation cover as long as possible during the year, it is necessary to promote culture associations. The rotations of "cereal-legume-forage" or "cereal-legume-cereal" are retained. For rangeland improvement, the vetch-oats, alfalfa, and clover present important opportunities for pastoral production and contribute significantly to soil protection.

- Steep slopes (higher than 15%): The results analysis shows that higher slopes are commonly used by cereal cultures that give low yields. Especially, in this case, the soil tillage in the direction of slopes causes ridges that eventually become water runoff channels (gullies) that quickly develop the gullies and ravines. Moreover, the tillage soils according to the slope direction increase erosion.

In conclusion, a sustainable soil management on steep slopes is necessary through the restoration of vegetation cover by the planting of multiple use species following the contours. This plantation technique must be combined with isohypse structures (benches, ditches, and cords) to conserve water and soil.

In the case of the Oued Beht watershed, fruit trees cultivation presents a promoter axis of the erosion control in the difficult terrain. The tree species proposed depend on agro-ecological areas. This operation needs also the consultation with the farmers concerned to choose trees species. Moreover, the olive, fig, and almond trees seem the most desired fruit trees by the population and the best adapted to the ecological conditions in the watershed. Second, the interline space will be used for the practice of the usual cultures respecting the principles of tillage soils following contour lines.

4.5.2 Rangeland management

The socio-economic study shows that the actual animal demand is high compared to production potential. Thus, the confrontation of the rangeland offers and livestock demand reveals an important deficit −31%.

The results analysis shows that the three livestock types (sheep, cattle, and goats) use rangelands intensively and continuously. Generally, the state of rangelands presents advanced degradation of vegetation resources. In addition, this usage mode is accentuated first by the severity of soil and climatic conditions which are often unfavorable and second by the nature land status that promotes non-rational exploitation of forage justified by its gratuity.

In this situation, the pastoral improvement is fully justified by the need to implement an intervention program to save the pastoral resources in the Oued Beht watershed. Thus, short-term actions are based particularly on the development and rational management of pastoral space, and then, in the medium term, the program can implement actions linked to improving driving livestock. Normally, the proposed actions tend to change the pastoralist habits and to support the incentive mechanisms related to fattening to reduce the pressure on the pastoral spaces.

- Rangeland users organization: The users organization into pastoral associations (or cooperatives) is a central action to be taken in parallel with the technical actions (plantation, closing and deferred grazing, and water point for livestock) in order to ensure sustainable use of rangelands. This organizational approach is the population interface with all partners to monitor actions and to defend the pastoral potential of the watershed.

- Deferred rotation grazing: In this case, the deferred rotation grazing is the technique used to enhance and restore the herbaceous and shrubs potential. It consists of prohibiting grazing in degraded areas in order to allow the natural regeneration with the development of herbaceous species richness and of forage quantity.

The duration of the deferred grazing depends on pastoral species. A short duration grazing is a rotation on 2–4 years, which is sufficient for the regeneration of herbaceous species and for the improvement of pastoral potential. However, the limitation of rights to use rangelands will be able to generate a forage imbalance that will directly increase the pressure on the surrounding lands and cause the accentuation of their degradation. Thus, to anticipate this problem, it is imperative to choose pastoral species with high nutrient supplies and to provide accompanying measures for population like compensation system linked to unexploited forage units and development of forage crops irrigated.

- Planting shrubs: In the case of the studied watershed, the survey analysis shows clearly that fodder shrubs are highly attractive to farmers. Thus, the shrubs present the advantage to provide their production in a late period of the year when other forage crops (including herbaceous vegetation) are low or zero. The introduction of tree plantations consists of soil tillage in the autumn before the first rains with the digging holes along the contour lines for planting shrubs and installing bleachers for planting cactus.

In conclusion, this technique aims to improve water balance and fight against erosion. The shrubs species that are recommended are Atriplex nummularia, *Medicago arborea*, Chamaecytisus albidus, and *Opuntia ficus-indica*. Moreover, the use of cactus plantations presents his pastoral role with the advantage of producing highly appreciated fruit that can provide substantial revenue for the users.

- Livestock watering points: The analysis of the surveys data shows that water shortage presents a major constraint, especially as the dry spells became frequent. In the summer, water resources become scarce and fail to cover the livestock needs. Several techniques for collecting and mobilization of water when they are available (in winters and flood periods) are proposed based on the watershed characteristics.

The proposed actions present great social utility and do not require large investments; they are adapted either to an individual or collective use. Thus, the actions include the preparation of water reservoirs, the capture from surface water sources, the development of existing wells, and the digging of new wells.

4.5.3 Forest management

Implementation of action plan linked to watershed forestry resource consists to restore degraded natural ecosystems (evergreen oak and thuya), which represent an economic and ecological importance. Thus, these actions aim to improve the vegetation cover, to protect the soil against erosion, and finally to halt the forest degradation.

- Forest rehabilitation: The biophysical analysis shows that watershed forests are located generally in difficult areas upstream. These ligneous formations have good adaptability and resistance to the negative impacts of climate and anthropogenic pressure. Thus, most of these forests suffer from a lack of natural regeneration.

Therefore, this difficult situation requires efforts in terms of natural regeneration with native species to ensure sustainability of these natural areas. Thus, the intervention program gives priority to the parties that have the potential for regeneration.

These actions are accompanied by water and soil conservation measures to reduce erosion and increase water storage capacity (step elements, benches, and terracing).

- Reforestation protection: The introduction of artificial plantations aims at the protection of degraded forests. Thus, the reforestation of denuded lands and badlands, with forest vocation, allows the soil protection, the runoff quality and quantity improvement, and production of wood products.

Considering the watershed bioclimatic conditions, the spectrum obtained from tree species proposed for reforestation is maritime pine, Aleppo pine, brutia pine, cypress, and eucalyptus trees.

- River system and badlands development: The hydrographic network is characterized by high density ratio of river and lakes; the soil losses are accentuated by this river system, and the erosion is generally active on soft to moderately vulnerable areas. This phenomenon is strongly observed in the central part of the watershed where the river system becomes increasingly ramified and individualized (**Figure 11**). This regressive evolution leads to a densification of ravines that can achieve the generalized gully erosion. This situation is clearly illustrated in the downstream part at the El Kansra dam. The sediment quantities resulting from this erosion are mainly transported downstream and contribute significantly to the dam siltation.

Finally, the proposed management strategy for vulnerable areas is based on a combination of two main actions: biological fixation and mechanical ravine correction. Thus, the two integrated actions stimulate vegetation installation and slope correction. The chosen technique combines the advantages, not only to limit sediment yield but also to promote the defense of infrastructure, good land, and public and private properties.

This socio-ecological development program gives special attention to aspects of social vulnerability, a major dimension of vulnerability to multiple factors including: low incomes, social exclusion, and natural hazards. Referring to this approach,

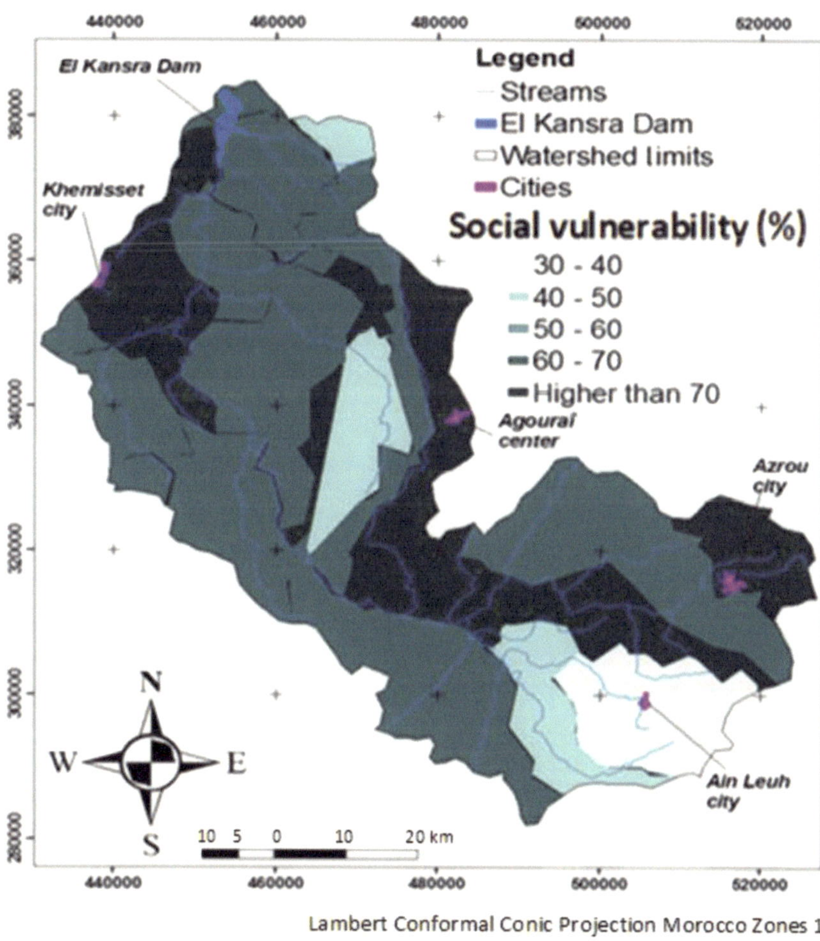

Figure 23.
Social vulnerability.

all people whose consumption expenditure is below the poverty line, which represents the minimum income considered adequate for each person, are considered vulnerable. In Morocco, on average, the poverty line is US $ 2.4 per person per day in rural areas [14, 15].

In fact, the survey design conducted in this study allowed us to exploit the income data of sampled individuals and to develop a simplified map representing, by homogeneous area, the percentage of individuals with an income below the minimum income deemed appropriate for each person (**Figure 23**).

4.6 Operational management program

The autocorrelation maps obtained (z-score) are used to delineate the priority interventions which correspond to the z-scores, statistically significant with values higher than 1.65 or less than −1.65 (**Figure 13**). Furthermore, the biological actions in degraded areas (by fruit plantation, regeneration, and reforestation) are materialized in spaces that express high spatial aggregation between soil loss and degraded vegetation cover. Consequently, the total area covered by this type of intervention

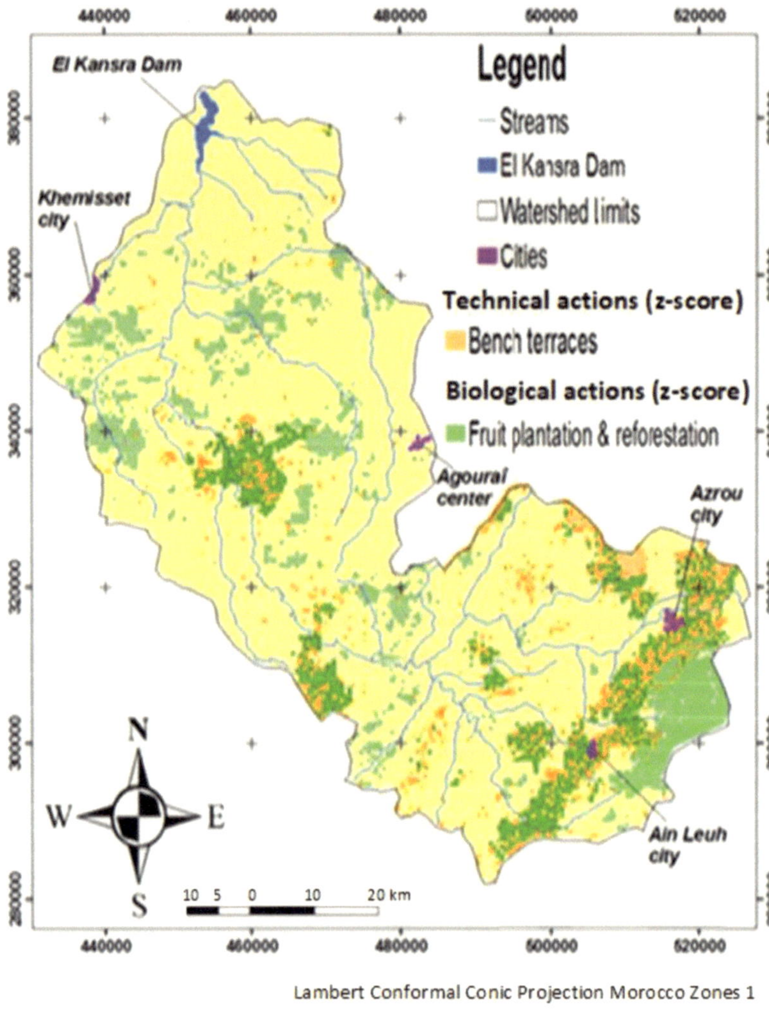

Figure 24.
Biological interventions.

is 87,351 ha, which represents 20% of the Oued Beht watershed. These biological interventions are concentrated mainly in the priority sub-catchments of Tigrigra, Ifrane, and Kharrouba (**Figure 24**).

On the other hand, the technical actions designed to reduce the slope effect consist in the establishment of benches, ditches, and terracing. These structures are programmed in high spatial aggregation between soil erosion and steep slopes. The total area covered by technical intervention is 22,753 ha, and a quarter of biological interventions is combined with technical measures, especially in the upstream part of the watershed (**Figure 24**).

In conclusion, the package of techniques of soil conservation and erosion control is developed in agricultural and sylvopastoral areas, starting from various types of soil tillage and vegetation cover, to different types of terraces, check dams, and stone bunds. Thus, the terracing is the selected agricultural technique for collecting surface runoff water, thus increasing infiltration and controlling water erosion used to transform landscape to steeped agrosystems in the mountainous regions (upstream).

5. Conclusion

This research paper proposes the development of a methodology analysis for soil erosion hazard and risk administration, especially a very few studies are dedicated to the mapping of soil loss risks. The use of analytical models based on space technology information processing has developed a GIS database on biophysical and topoclimatic parameters in Oued Beht watershed. Thus, the procedure described evaluates the soil loss risk and siltation of El Kansra dam, located in the upstream side.

The present study has implemented a cartographic approach based on the integration of spatial remote sensing tools (GIS) and spatial analysis functionalities linked to the initial state of the studied watershed. Thus, the central objective is to define the guidelines of the strategic spatial planning dedicated to erosion risk management. Moreover, although some studies have combined biophysical data and the constraints identified in the socio-economic analysis in order to understand the conditions of water erosion, they generally do not consider the statistical autocorrelation to develop strategy for priority management of watersheds. In this perspective, the cartographic restitution of spatial clusters obtained identifies priority areas and establishes the first interventions across the watershed.

The results obtained from the spatial autocorrelation analysis concerning socio-ecological components show that the priority actions are needed for almost 20% of the Oued Beht watershed. Thus, all priority areas identified are affected by the biological techniques (fruit plantation, regeneration, and reforestation in adequate slopes) that mitigate the factor, which expresses the lack of vegetation cover.

In addition to that, the spatial aggregation map shows also that the appropriate soil conservation practices (terracing) correspond to a quarter (15%) of the priority areas. Thus, this category of intervention aims to reduce the negative effects of the topographic factor with the establishment of terracing structures (**Figure 13**). The main purpose of the terracing application is to improve the usefulness of steep slopes and to increase their agricultural potential. This function is realized by creating the level surfaces according to contour lines of transformed slopes. The level, bench platform allows spreading the surface runoff water, decreases its speed, and thus allows more time for water infiltration into soil profile.

In conclusion, this approach has allowed developing a planning program with successful techniques for soil erosion control in degraded areas linked to steep slopes, climatic conditions, and erodible soils.

It is obvious that this approach, based on ground measurements combined with geographic information systems, must be accompanied by a regular monitoring system by updating continuously the part of the spatial model derived from remote sensing. Furthermore, the stable part of the geospatial database consists of intrinsic factors (lithology, soil, drainage density, etc.) and the dynamic part to control includes biotic factors related to the soil occupation and needs evolution of the local population.

Although this analysis was conducted to the master plan of watershed development and has identified environmental constraints (soil and water degradation) characterizing priority areas, it is necessary to refine this analysis through a participatory action plan. Thus, this zonal analysis will specify for each year the interventions to be implemented and the financial package, by considering the needs and perspectives of the rural population. Moreover, the effectiveness of the proposed techniques can be limited especially if the local population is opposed or, in some cases, found to be expensive to build and maintain.

Finally, this research work demonstrates the potential and merits of spatial analysis techniques to evaluate the erosion risks. An indicative mapping designed for the management and risk prevention is obtained, to control the source and quality of input and to characterize the conditions of validity of the models. However, the difficulties encountered in the collection of quantitative damage data, usually, due to the lack of historical information, refer to the idea that it would be necessary to create an observatory and full database related to water erosion damage. Thus, research is needed to introduce also the temporal component (probability of erosion and return period) in a decision support perspective to implement a regional sustainable planning.

Acknowledgements

This research paper was partially supported by the Identification and Modeling Laboratory of Natural Environment (LIMEN), Mohammadia School of Engineers in Mohammed V University (Morocco). I would also like to express my gratitude to some of my colleagues who were generous in providing guidance, and without their help, this project could not have been accomplished.

Dedication

I dedicate this paper to my parents Ahmed and Jamila, my sisters and my nephew Mouad, my dear family, and to the soul of my uncle Hassan Touissate. I also dedicate this modest work to my dear wife Ihsane, friends, and colleagues, and without their encouragement, I could not have written this.

Author details

Rabii El Gaatib
Ministry of Agriculture, Fisheries, Rural Development, Water and Forests, Rabat, Morocco

*Address all correspondence to: rabii_elgaatib@yahoo.fr

References

[1] HCWFFAD High Commission for Water, Forest and Fight Against Desertification. National Watershed Management Plan. Rabat; 1996

[2] HCP High Commission for Planning. Socio-economic and Demographic Characteristics of the Population Based on the General Census Population and Housing. Rabat; 1994. Available from: http://bibliotheque.insee.net/index.php?lvl=notice_display&id=147899 [Accessed: August 15, 2016]

[3] HCP High Commission for Planning. General Census of Population and Housing; 2004. Available from: www.hcp.ma/file/111366 [Accessed: August 21, 2016]

[4] HCP High Commission for Planning. Moroccan Statistical Yearbook; 2011. Available from: http://www.hcp.ma/downloads/Annuaires-statistiques-regionaux_t11956.html [Accessed: June 18, 2016]

[5] WBAS Water Basin Agency of Sebou ABHS. Integrated Water Resource Management Plan in the Sebou Watershed. Rabat; 2011. Available from: http://www.abhsebou.ma/wp-content/uploads/2018/01/Pdaire-sebou-22-10-2015-.doc [Accessed: August 10, 2016]

[6] Getis A, Ord JK. Analysis of Spatial Association by Use of Distance Statistics, Geographical Analysis. 1992; pp. 189-206. Available from: www.archive.nefmc.org/tech/.../Getis-Ord%20 statistic.pdf [Accessed: August 10, 2016]

[7] Roose E. Erosion and Runoff in West Africa: Twenty Years of Measurements in Small Areas, Works and documents ORSTOM no. 78. Paris; 1977. ISBN 2-7099-0480-2. Available from: http://horizon.documentation.ird.fr/exl-doc/pleins_textes/pleins_textes_6/Tra_d_cm/09011.pdf [Accessed: August 11, 2016]

[8] Roose E. Soil and water conservation lessons from steep slopes farming in French speaking countries of Africa. In: Moldenhauer WC, Hudson NW, editors. Conservation Farming on Steep Lands, Soil Conservation Society of America, Ankeny, Iowa; 1988. pp. 129-139. DOI: 10.1017/S0889189300003489

[9] Wischmeier WH, Smith DD. Prediction rainfall erosion losses from cropland east of the Rocky Mountains: A guide for selection of practices for soil and water conservation. U.S. Department of Agriculture, Washington DC; 1965

[10] Wischmeier W.H., Smith D.D. Prediction rainfall erosion losses, a guide to conservation planning science. U.S. Department of Agriculture, Washington DC; 1978

[11] Francou J, Rodier JA. Classification Test of Maximum Floods Observed in the World. In: Cah. ORSTOM. ser. Hydrol, vol. IV(3). Paris; 1967. Available from: http://horizon.documentation.ird.fr/exl- doc/pleins_textes/pleins_textes_4/hydrologie/14846.pdf [Accessed: August 11, 2016]

[12] Guillot P, Duband D. The GRADEX Method for Calculating the Flood Probability from Rainfall, SHF, 10th Hydraulics Day, Question 1, Report 7, Paris; 1968. In: Guillot P, La Houille Blanche, N°3, 1971. pp. 209-218. DOI: 10.1051/lhb/1971014

[13] MMAMF Moroccan Ministry of Agriculture and Marine Fisheries. General Census of Agriculture, Rabat; 1996. In: MMAMF. Atlas of Moroccan agriculture, Rabat; 2018. Available from: http://www.agriculture.gov.ma/sites/default/files/ATLASsynthese.pdf

[14] HCP High Commission for Planning. Evolution of Living Standards, Inequality and Poverty in Morocco; 2009. Available from: www.hcp.ma/file/111826 [Accessed: June 05, 2016]

[15] HCP High Commission for Planning. Socio-Demographic Indicators; 2010. Available from: www.hcp.ma/file/103089 [Accessed: June 05, 2016]

[16] High Commission for Planning. Regions of Morocco; 2010. Available from: www.hcp.ma/file/129637 [Accessed: June 05, 2016]

[17] Rabarimanana M, Andriamasimanana R, Rasolomanana E, Robison L. Study of Vulnerability in Ihotry Watershed to Sheet Erosion. Madamines, ISSN 2220-0681, Vol. 4; 2012. 55p. Available from: https://www.scribd.com/document/362612713/Madamines4-5 [Accessed: August 12, 2016]

[18] ESRI. ArcGIS Desktop: Tool Reference, Spatial Statistics Toolbox; 2014. Available from: http://help.arcgis.com/fr/arcgisdesktop/10.0/help/005p/005p00000006000000.htm [Accessed: March 08, 2016]

[19] Emberger L. Climate Biogeographic Classification. Vol. 7. Collection of Botanical Geological and Zoological Laboratories Works, Montpellier Faculty of Sciences. 1945;7:3-43

[20] Rango A, Arnoldus HMJ Watershed Management. In: FAO Technical Manual; 1987. pp. 1-11

[21] El Gaatib R, Larabi A. Soil erosion effects on the natural resources of watersheds and the siltation condition of dam reservoirs: Application to OuedBeht watershed upstream of El Kansra dam (Morocco). Géo Observateur No. 21. Rabat; 2014:35-45. Available from: https://ecitydoc.com/download/geo-observateur-n-21-centre-royal-de-teledetection-spatiale_pdf

[22] El Gaatib R, Larabi A. Integrated evaluation of soil erosion hazard and risk management in the Oued Beht watershed using remote sensing and GIS techniques: Impacts on El Kansra dam siltation (Morocco). Journal of Geographic Information System. 2014;**6**:677-689. DOI: 10.4236/jgis.2014.66056

[23] HydroEurope. Final Engineering Report. 2012. pp. 20-32. Available from: http://www.hydroeurope-team1.jennylestil.com/FinalEngineeringReport.pdf?forcedownload

[24] Roose E. Use of the universal soil loss equation to predict erosion in West Africa. In: Soil Erosion: Prediction and Control, Soil Conservation Society of America, Ankeny, Iowa, Special Publication No. 21; 1976. pp. 60-74. Available from: https://core.ac.uk/download/pdf/39881180.pdf [Accessed: August 02, 2016]

[25] Roose E. Conservation des sols en zone méditarrénenne. Synthèse et proposition d'une nouvelle stratégie de lutte antiérosive: La Gestion et Conservation des Eaux et Sols (GCES). Centre ORSTOM Série Pédologie, Montpellier. 1991;**26**:145-181